Mathematical Methods
Basic Math

By J. R. Miller

ISBN 978-0-9801487-5-6

Printed in the United States of America

Cover image: NASA photograph of Earth from the Apollo 17 mission set against the background of the reflection nebula. Located 1,000 light-years from Earth in the constellation Perseus, the reflection nebula NGC 1333 epitomizes the beautiful chaos of a dense group of stars being born. The stars are beginning to move away from their formative cloud, seen in red and green. Jets can be seen coming off the young stars as they make their way into the cosmos.

Mathematical Methods
Basic Mathematics

Table of Contents

Introduction

Mathematical Methods was developed to enable people to master mathematics using a systematic approach. This is to be accomplished by understanding the definitions, history, applications, and algorithms (methods for problem solving) of the various topics of mathematics. Most books and resources on mathematics give you a formal mathematical definition and then show one or two worked examples. Some textbooks go further and provide historical notes and show some of the steps required to solve particular problems. In most worked examples, one or more steps are omitted and the student is left to try to figure out how the final answer was derived. Other textbooks give you applications and examples but leave out how to approach solving different classes of problems.

The goal of *Mathematical Methods* is to include everything you need to know to master each level of mathematical studies. The format for all topics is as follows:

Topic Definition
Topic History
Topic Applications
Topic Rules, Theorems, and Proofs
Topic Algorithm (the method used for solving a class of problems)
Topic Examples

For all major topics, a definition in non-technical form is provided. You have to be able to define something before you can understand what it is or how to work on it. Where applicable, the definition is followed by a brief history of the topic, which provides information for understanding the origins of mathematical concepts. The history is followed by a brief overview of the applications of the topic. Understanding what the mathematical concepts can be used for helps maintain interest in the subject and assists you in remembering why you need to learn the material in the first place.

The applications are followed by the formal rules, theorems, and proofs. These are the formal definitions that you find in any standard textbook. This is followed by the algorithms (a step-by-step procedure for solving a problem) that are used to solve the type of problem under consideration. Finally, there are worked examples of the typical types of problems that you will see for each topic with no steps left out and each step explained. The general approach taken by *Mathematical Methods* is as follows:

1. You see symbols, letters, and numbers. To see is to perceive by mental vision, and form an idea or conception.
2. You make a determination of what form the symbols and numbers represent and if there is a pattern. You determine the relations of the form and what they reveal. You define the form and its attributes.
3. Once you determine what form the pattern represents you mentally place it into the appropriate class or type.
4. Finally, you apply the algorithm for the solution to the indentified class of problem.

A simple example may help to clarify these steps. When you see an expression in the form of $\frac{8}{4}$ what do you observe? You see two numbers divided by a line. This form represents a fraction.

Until you recognize this type of fraction you may ask the following questions. What type of fraction is it? What are its properties? What pattern does it follow? What are the rules of fractions and which rules pertain to this particular fraction? Once you have determined the answers to these questions it is simply a matter of applying the rules and algorithms for solving the problem. In this case, you have an improper fraction, which can be reduced to a value of 2.

Until you understand what the symbols, letters, and numbers mean, know what the forms or patterns represent, know the definitions of the concepts and what types of operations can be performed, and know the rules and algorithms for solving a particular type of problem, the symbols are meaningless gibberish.

Learning mathematics is like learning a second language. To become proficient in a second language you have to speak the language, to become proficient in mathematics you have to solve problems. Deliberate practice solving large numbers of mathematical problems with pencil and paper is how you get good at mathematics. As a mathematics professor once stated, math is a motor skill. Fine motor skill involves deliberate and controlled movements requiring both muscle development and maturation of the central nervous system. Just as children develop fine motor skills as they age, math skills are developed though the use of eye-hand-brain coordination. You cannot become skilled in mathematics just by reading mathematics textbooks and watching other people work problems.

Repetition and practice are the only way to get good at any endeavor. Scientific studies have consistently shown that there is no evidence of high-level performance without experience and practice. Even the most accomplished people need around ten years of hard work (the ten year rule) before they reach world-class levels. The best people in any field are those who devote the most hours to consistent constant deliberate practice. Deliberate practice is an activity that is explicitly intended to improve performance and involves high levels of repetition, constantly working for objectives just beyond one's level of competence, measuring performance, and providing feedback for further improvement.

With mathematics it is essential that you master the basics in each topic before moving on to the next. This is because mathematics is truly a foundational science. For example, you cannot understand calculus until you have mastered algebra and trigonometry. Without a complete understanding and mastery of lower level mathematics, you don't have a chance understanding higher level mathematics.

When children first start studying math and they have to learn addition and subtraction, then multiplication and division, most kids start asking, "Why do I have to learn this stuff?" They probably got a response something like; "You have to learn your arithmetic so you can advance to the next grade." Well, that really isn't the reason. The real reason you need to learn math is because if you want to succeed in modern society you have to know mathematics.

Expertise in mathematics provides the basis for careers in many of the fastest growing fields including actuarial sciences, engineering, statistics, bio-informatics, quantitative research in equities and derivatives, computer science, and math instruction. In the relatively new field of mathematical finance, people with a PhD. in mathematics first-year compensation can range between \$100,000 and \$300,000, or more depending on background and experience. Contrary to popular perception, mathematics is the key to high paying jobs.

Mathematics is essential for a complete understanding of the world, and is critical to success in an information and technology based society. Although some topics in mathematics can be tedious and require hard work, your efforts will be richly rewarded. Once you learn advanced mathematics you can learn and understand any other field of study. Do yourself and your children a favor, and study mathematics. It will not only help your career it will help your brain. To help get you started and to show the importance of mathematics in the development of today's modern technological society a brief history of mathematics is provided below.

Mathematics

Definition: Mathematics is the collective name applied to all those sciences in which operations in logic are used to study the relationship between quantity, space, time, and magnitude. Mathematics employs a special kind of language using symbols, numerals, and letters. Their use is determined by the rules of logic within each of the fields of study and serves as a means of abbreviation, both in thinking, and in visual representation.

History: Mathematics arose from the need to catalog and count property and people, and the practical needs of agriculture and trade. It seems reasonable that the earliest humans used fingers to count and communicate numbers and the widespread use of the decimal system is the result of the fact that humans happen to have ten fingers and ten toes. Decimal refers to the base ten number system, which uses the symbols 0 through 9 to represent values.

The first record of the use of mathematics beyond mere counting comes from the Egyptians and Babylonians. Building on the concept of counting, arithmetic was developed. Today, arithmetic refers to numerical computation involving the four fundamental operations of addition, subtraction, multiplication and division, which are performed in accordance with the axioms for real numbers.

Most of our knowledge of Egyptian mathematics comes from the Rhind papyrus written by the scribe Ahmes about 1650 B.C., and the Moscow papyrus. Papyrus is an early form of paper made from the papyrus plant. The Rhind Papyrus was named after the Scottish Egyptologist Alexander Henry Rhind, who purchased it in 1858. The Rhind papyrus, which now resides in the British Museum, consisted of a collection of practical mathematical problems and solutions. Many of the problems were in the form of what we would today call "story problems." The Rhind papyrus also includes several tables of fractions. The Moscow Papyrus has several of the same type of problems and solutions found in the Rhind papyrus. It also includes a description of the truncated pyramid. A truncated pyramid has a square base, which narrows to a square top. In other words, a normal pyramid with the top cut off. A truncated pyramid, also known as a frustum, is created by slicing the top off with a cut made parallel to the base.

The Babylonians used the cuneiform system of writing. Cuneiform used a reed to impress wedge-shaped marks onto the surface of clay tablets. These cuneiform symbols were used to create tables to aid calculation. Babylonian mathematics included the development of a sexagesimal (base 60) positional notation number system and the use of the Pythagorean Theorem many years before its formal proof. Sexagesimal measure is the degree system of measure in which a complete revolution is considered to be 360 degrees. The unit of measure, or 1 degree, is equal to 1/360 of a complete revolution. The degree is divided into 60 minutes (60') and the minute into 60 seconds (60").

The innovation of positional notation streamlined the process of representing numbers and kept the number of symbols to a minimum. By contrast the Roman numeral system, which was not a place value system, required a large number of symbols, such as I, V, X, L, C, D, M, and was very cumbersome to work with.

In the 6th century B.C., the Greeks, including Pythagoras, Euclid, Archimedes, and Thales transformed arithmetic and geometry into systems of logic. The Greek Period produced some of the greatest mathematicians and philosophers the world has known. Greek mathematicians developed the foundations of modern mathematics, and had a profound influence on philosophy and scientific method. Greek geometry in particular stands as one of the greatest achievements of the human intellect.

During the Greek period mathematics advanced from arithmetic to abstraction. To abstract is to consider apart from application to or association with a particular instance. Abstraction in mathematics is the process of extracting the underlying essence of a mathematical concept, removing any dependence on real world objects with which it might originally have been connected, and generalizing it so that it has wider applications.

In the Middle Ages Arabic mathematicians, through the abstraction of arithmetic, developed algebra and trigonometry. Then in the 19th century, Rene Descartes combined algebra and geometry to create analytic geometry. Analytic geometry provided the foundation for new disciplines in higher mathematics including calculus, which was developed independently by Isaac Newton and Gottfried Leibniz. The development of modern mathematics continued with probability theory, non-Euclidean geometry, topology, set theory, game theory, and fractal geometry.

Many areas of mathematics began with the study of real world problems, before the underlying rules and concepts were identified and defined as abstract structures. For example, geometry has its origins in the calculation of distances and areas in the real world; statistics has its origins in the calculation of probabilities in gambling; and algebra started with methods of solving problems in arithmetic.

Abstraction is an ongoing process in mathematics and the historical development of many mathematical topics exhibits a progression from the concrete to the abstract. As mathematics becomes more advanced it also becomes more abstract. The first steps in the abstraction of geometry were made by the ancient Greeks, with Euclid being the first person to document the axioms of plane geometry.

Mathematics has evolved to become a large and diverse field of study. An overview of the branches and classification of the various fields of mathematics is included in the appendix.

Applications: Mathematics is used in almost every field of study, from art with the geometry of perspective, to political science and the mathematics of voting and game theory. Mathematics is essential for the study of business, operations research, and economics. There are many fields that cannot be mastered without advanced mathematics including physics, chemistry, engineering, computer science, and biology.

Numbers

Definition: A **number** is a concept of quantity which could be a single unit or a collection of units. The value of the quantity is represented by a symbol called a numeral. **Numeration** is the process of writing or stating numbers in their natural order. Numbers can be classified according to the types of units in the collection. These classes of numbers include the natural numbers, whole numbers, integers, rational numbers, and irrational numbers. These classes are all part of the set (collection of objects) of real numbers. An additional class of numbers known as complex numbers or imaginary numbers will be defined in *Mathematical Methods Algebra*.

Zero is a unique unit and will be examined first. It is the first number in the collection or set of whole numbers. Zero fulfills a central role in mathematics as the **additive identity** (if you add any number and zero you get back the original number) of the integers, real numbers, and many other algebraic structures. As a **digit** (one of the symbols of a number system), 0 is used as a placeholder in place value systems. Zero is an even number. Zero is neither positive nor negative, and is neither a prime number nor a composite number. Division by zero is not defined. Zero is the number that separates the positive numbers from the negative numbers.

The exact origin of zero is uncertain. The Babylonians used a character for the absence of a number and did use a place value system. It is believed that the Greeks also had an understanding of place value. The Greek *lacuna* (nothing) symbolized by the Greek letter omicron (o) or the Hindu small circle ($^{\circ}$) could be the origin of the form of zero used today. The current consensus is that zero appeared in India in the 9th century and that it is most likely a Hindu invention.

Classes of Numbers

The main classifications of numbers are given below. There are many other categories or classes of numbers most of which are used in higher level mathematics and number theory and will not be reviewed at this time.

Natural or counting numbers - The numbers 1, 2, 3, 4, ... are the counting numbers. The three periods or dots after the 4 are referred to as an **ellipsis**, which indicates an omission, and is used in mathematics to mean continue on in a like manner, continue this pattern, or that the pattern continues without end. The natural numbers are called counting numbers because each counting number can be used to count physical objects.

Whole numbers - The whole numbers 0, 1, 2, 3, 4, ... consist of the counting numbers together with the number zero.

Integers – The integers ... ,-2, -1, 0 1, 2, ... consist of the whole numbers or positive integers, and the negative integers. The positive and negative integers are also referred to as signed numbers. A **signed number** is a number proceeded by a plus (+) or minus (-) sign indicating a positive (+) or negative (-) number. Signed numbers are also called directed numbers since the sign can indicate the position or direction of a number, relative to zero on a number line. The negative numbers are to the left of zero on the number line and the positive numbers are to the right of zero on the number line. As you go further to the left on the number line towards negative infinity ($-\infty$) the numbers get smaller. As you go further to the right on the number line towards positive infinite (∞) the numbers get larger. **Infinite** is something that is limitless (without any limit) or endless in space, time, extent, or size; or becoming large beyond any fixed bound. **Infinity** is the state or quality of being infinite. John Wallis is credited with introducing the infinity symbol (∞), in 1655. The infinity symbol is a **lemniscate**, which is a figure-eight shaped curve. The number line is examined in more detail in the section on the number line and coordinate systems.

Rational numbers – A rational number is a number that can be expressed as a ratio of two integers such as $\frac{2}{3}$ or $\frac{3}{4}$. Since a number such as 5 may be written as $\frac{5}{1}$, all integers are rational numbers. Rational numbers are either repeating or terminating decimals such as $\frac{1}{3}$ (which equals .333...).

Irrational numbers – An irrational number is a real number which cannot be expressed as the ratio of two integers. Irrational numbers such as π (3.1415...), $\sqrt{2}$ (1.4142...), and $\sqrt{3}$ (1.7320...) are infinite decimals that are non-repeating and non-terminating.

Real Numbers – The set of real numbers includes zero, the positive and negative rational numbers, and the positive and negative irrational numbers. They are called real numbers to distinguish them from imaginary numbers.

Numbers are further categorized as even and odd numbers, and prime and composite numbers. **Even numbers** are any number that is exactly divisible by two. **Odd numbers** are numbers which are not even and therefore not divisible by two. **Prime numbers** are numbers that can only be divided by 1 and the number itself. Zero and 1 are not prime numbers. The only even prime number is 2. For example, 17 is a prime number because it can only be evenly divided by 1 and 17. The number 21 is not a prime number because it can be evenly divided by numbers other than 1 and itself, in this case the numbers 3 and 7. The first 25 prime numbers are 2, 3, 5, 7, 11, 13, 17, 19, 23, 29, 31, 37, 41, 43, 47, 53, 59, 61, 67, 71, 73, 79, 83, 89, and 97. **Composite numbers** are numbers that have two or more prime factors. A **factor** is each of the numbers or letters multiplied together to obtain a given product. Factors will be reviewed in greater detail in the section on fractions.

Place Value

Place value, or positional notation, is a numeration system in which a real number is represented by an ordered set of characters or numbers where the value of a character or number depends on its position. Each position is related to the next by a constant multiplier called the base of that numeral system. A **base** is the number of different digits that a system of counting uses to represent numbers. For example, base ten or the decimal system, has ten digits, the digits 0 to 9. The resultant value of each position is the value of its symbol or symbols multiplied by a power of the base. The total value of a positional number is the sum of the resultant values of all positions. Positional notation is discussed in detail in the section on number systems.

In the base ten number system, the first place or position to the left of decimal point from right to left, is the units digit (the digits 0 to 9), the second place is the tens digit, the third place is the hundred's digit, and so on, when necessary zero is used as a placeholder. The number 364 means 3 hundreds, plus 6 tens, plus 4 units, or 300 + 60 + 4. As we will see later, place value is important in number systems, scientific notation, and algebra. The figure below displays the place values for the number 3,305,237,461.53731901.

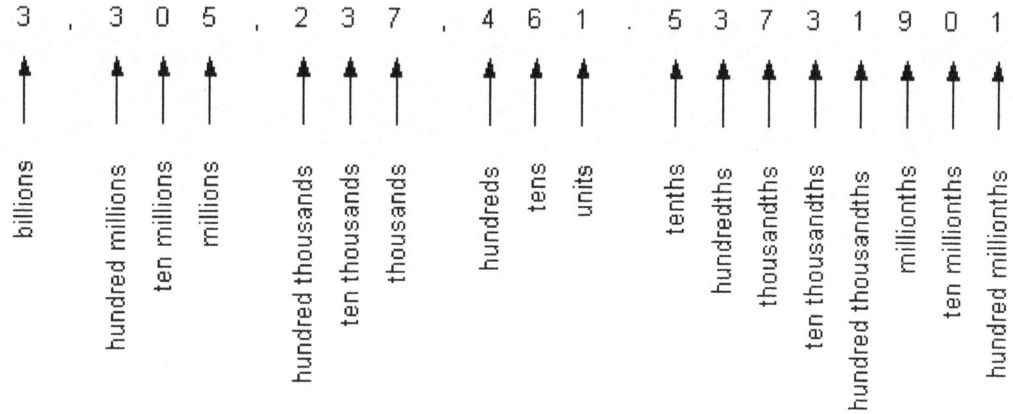

Summary

Number is a concept of quantity that could be a single unit or a collection of units. Zero is a symbol that is used as a placeholder in numeral systems using positional notation and is used as an additive identity.

The classification of numbers includes the natural or counting numbers, which are called counting numbers because they can be used to count physical objects; whole numbers, which consist of the counting numbers together with the number zero; integers, which consist of the whole numbers or positive integers, and the negative integers; rational numbers, which are numbers that can be expressed as a ratio of two integers, and irrational numbers, which are numbers that cannot be expressed as the ratio of two integers. The set of real numbers includes zero, the natural numbers, the whole numbers, the positive and negative integers, the rational numbers, and the positive and negative irrational numbers.

Place value is a numeration system where the value of a character or number depends on its position. When the place or position changes the value of the number changes as well. The number 5 represents five units but if the 5 is shifted left one position and a 5 is placed in its previous position giving 55, the face value is still 5 but the number now represents 5 tens or 5 x 10 = 50 plus 5 units, which is 55. The number 5 still represents five of something but if the 5 is in the tens position the value is 50, if the 5 is in the hundreds position the value is 5 x 100 = 500 and so on.

Arithmetic

Now that number, zero, and the classes of numbers have been defined, it is time to learn how to perform operations on numbers. In mathematics, there are operations and formulas you will have to memorize in order to perform calculations in a timely manner. The operations of arithmetic require memorization. You can use the arithmetic tables in the appendix to help you. You should memorize the addition, subtraction, multiplication, and division tables up to the number twelve as shown in the tables.

Definition: Arithmetic involves the study of quantity, especially as the result of combining numbers and is formally defined as the science of number and computation. The basic operations of arithmetic are addition, subtraction, multiplication, and division.

Applications: Basic arithmetic is used in everyday life for tasks such as balancing a checkbook, verifying charges on invoices, calculating taxes, and determining payments on automobile loans and home mortgages.

The operations of arithmetic for small numbers can be visualized by visualizing physical objects. For addition, which is the grouping of objects, you can visualize adding 3 and 6 as having two baskets of apples. In one basket there are 3 apples and in the other basket there are 6 apples. If you move the 3 apples from the one basket and place them in the basket with 6 apples you now have 9 apples. You have just added 3 and 6, which is equal to 9. Subtraction is the inverse or opposite of addition. If you start with your basket of 9 apples and remove 3, you will have 6 remaining. You have just subtracted 3 from 9, which is equal to 6.

Multiplication is repeated addition. To visualize multiplication we start with 3 baskets with each basket containing 3 apples. If you combine all three baskets into one empty basket you are essentially doing repeated addition. You are adding 3 apples three times for a total of 9 apples. You have just multiplied 3 times 3, which is equal to 9. Division is the inverse or opposite of multiplication or repeated subtraction. If you start with the basket of 9 apples and you remove 3, you have 6 remaining. If you remove another 3, you will have 3 remaining and you will be back to where you started, with 3 baskets of 3 apples. You have just divided 9 by 3, which is equal to 3.

Addition

Definition: Addition is the process of finding the number that is equal to two or more numbers grouped together. For example, if you have five apples and someone gives you three more, you have a total of eight apples. The two numbers (addends) are 5 and 3, and when grouped together or summed they equal 8. The terms of addition are the addends or the numbers to be added and the sum, which is the result, obtained by combining the addends. The example below can be read as 5 plus 3 equals 8, or the sum of 5 and 3 is 8.

$$\begin{array}{r} 5 \\ +3 \\ \hline 8 \end{array}, \quad \begin{array}{c} \text{addend} \\ \text{addend} \\ \hline \text{sum} \end{array}$$

Algorithm: Simple addition of single digits is a matter of memorization. The general rule for addition is as follows:

1. Write the addends under each other, units under units, tens under tens, etc.
2. Begin at the right and add the unit's column. Write down the units digit of the sum and carry the tens value to the next column which is the tens place.
3. Add the ten's column including the tens value from the units column. Write down the digit representing the number of tens and carry the hundreds digit to the hundred's column. Continue in the same manner until all columns have been summed.

There are two methods for adding columns of large numbers, the **carrying method**, as described above, and the **partial sums method**. For addition using the carrying method, write the numbers to be added so that the units are lined up under units, tens under tens, hundreds under hundreds and so on. Begin at the top right and add the unit's column. Write down the units digit of the sum and if the sum is greater than 9 carry (add to) the tens digit to the next column, which represents the place value for ten. Add the ten's column and write down the ten's digit and carry the hundreds digit to the hundreds column. When adding numbers with the same sign (either positive or negative) add the numbers and keep the same sign. When adding numbers with different signs subtract the numbers and keep the sign from the larger number.

Example 1: Find the sum of 2134 + 4742 + 7358 + 3972.

$$\begin{array}{r} 2134 \\ 4742 \\ 7358 \\ +3972 \\ \hline 18206 \end{array}$$

The sum of the unit's column is 16 (4 + 2 + 8 +2 = 16). Write 6 in the unit's place and carry the 1 to the ten's column. The sum of the ten's column is 19 (3 + 4 + 5 + 7 = 19) plus the 1 carried over from the unit's column which equals 20. Write a 0 in the ten's place and carry the 2 to the hundred's column. The sum of the hundred's column is 20 (1 + 7 + 3 + 9 = 20) plus the 2 carried over from the ten's column which equals 22. Write a 2 in the hundred's column and carry the 2 to the thousands column. The sum of the thousand's column with the 2 carried over from the hundred's column is 18 (2 + 4 + 7 + 3 + 2 = 18) and you have the final answer as shown below.

Example 2: For the partial sums method, add each column separately. Write one sum under the other, moving the place values over one space to the left. Then add the columns.

2134	
4742	
7358	
+3972	
16	Sum of units
19	Sum of tens
20	Sum of hundreds
16	Sum of thousands
18206	Sum of addends

Example 3: When adding numbers with the same sign (either positive or negative) add the numbers and keep the same sign.

$$\begin{array}{r} 5 \\ +3 \\ \hline 8 \end{array} \qquad \begin{array}{r} -5 \\ +-3 \\ \hline -8 \end{array}$$

Example 4: When adding numbers with different signs subtract the numbers and keep the sign from the larger number.

$$\begin{array}{r} 5 \\ +\ -3 \\ \hline 2 \end{array} \qquad \begin{array}{r} -5 \\ +\ \ 3 \\ \hline -2 \end{array}$$

Numbers can be added horizontally $5+3=8$ as well as vertically $\begin{array}{r} 5 \\ +3 \\ \hline 8 \end{array}$.

Subtraction

Definition: Subtraction is the process of finding a quantity which when added to one of two given quantities will give the other. The terms of subtraction are the minuend or the number to be reduced; the subtrahend or the number to subtract; and the remainder or difference. Subtraction is the inverse (opposite in order) of addition. Using the example from addition above if you have eight apples and you give three away you have five remaining. The example below can be read as 8 minus 3 equals 5, or the difference of 8 and 3 is 5.

$$\begin{array}{r} 8 \\ -3 \\ \hline 5 \end{array}, \quad \begin{array}{c} \text{minuend} \\ \text{subtrahend} \\ \hline \text{remainder or difference} \end{array}$$

Algorithm: Subtraction of single digits, like addition, is a matter of memorization. The general rule for subtraction is as follows:

1. Write the subtrahend under the minuend, units under units, tens under tens, etc.
2. Begin at the right and subtract each number of the subtrahend from the corresponding number of the minuend and write the remainder or difference underneath.
3. If any number of the subtrahend is greater than the minuend, increase the minuend using the exchange method (see below for more on the exchange method).

For larger numbers, if the digits of the minuend are larger than the corresponding digits of the subtrahend you simply take the smaller number from the larger. If some of the digits of the minuend are smaller than the corresponding digits of the subtrahend, the exchange method is required. To exchange is to take one thing in return for another which is regarded as an equivalent. With the exchange method, starting from the right and moving to the left, you exchange or borrow from one place value and use the value in another place value. This increases the value of the minuend so that is larger than the subtrahend and then you can take the smaller number from the larger.

Example 5: To subtract positive and / or negative numbers, change the sign of the number being subtracted (the subtrahend) and then add.

a. $\begin{array}{r} 5 \\ -+3 \\ \hline 2 \end{array} \rightarrow \begin{array}{r} 5 \\ +-3 \\ \hline 2 \end{array}$ The sign of 3 is positive. To subtract, change the sign of 3 to -3 and add.

$$\text{b.}\quad \begin{array}{r} -5 \\ --3 \\ \hline -2 \end{array} \rightarrow \begin{array}{r} -5 \\ +\ 3 \\ \hline -2 \end{array}$$

The sign of 3 is negative. To subtract, change the sign of -3 to 3 and add.

$$\text{c.}\quad \begin{array}{r} -5 \\ -+3 \\ \hline -8 \end{array} \rightarrow \begin{array}{r} -5 \\ +\ -3 \\ \hline -8 \end{array}$$

The sign of 3 is positive. To subtract, change the sign of 3 to -3 and add.

$$\text{d.}\quad \begin{array}{r} 5 \\ --3 \\ \hline 8 \end{array} \rightarrow \begin{array}{r} 5 \\ +\ +3 \\ \hline 8 \end{array}$$

The sign of 3 is negative. To subtract, change the sign of -3 to +3 and add.

An easy way to remember how to subtract a negative number is to remember that negative times a negative is a positive. In example 5d above $-(-3)$ is positive.

Example 6: The digits of the minuend are larger than the corresponding digits of the subtrahend. Starting with the right column (unit's place) subtract 6 from 9 which equals 3. From the ten's place subtract 2 from 4 which equals 2, and finally subtract 5 from 7 which equals 2 giving you a final answer of 223.

$$\begin{array}{r} 749 \\ -526 \\ \hline 223 \end{array}$$

Example 7: Some of the digits of the minuend are smaller than the corresponding digits of the subtrahend. Using the exchange method, exchange 1 ten for 10 units. This exchange results in the 3 in the tens column becoming 2 (3 – 1 = 2) and the 4 in the units column becoming 14 (10 + 4 = 14). Subtracting 5 from 14 equals 9. Since the 6 in the tens column is less than 2 exchange 1 hundred for 10 tens which are added to the 2 tens resulting in 12. Subtracting 6 from 12 equals 6. Next, exchange 1 thousand for 10 hundreds and add to the 4 in the hundreds column and subtract 9 hundreds from 14 hundreds. Subtracting 9 from 14 equals 5. Finally, subtract 1 thousand from 3 thousand. The final answer is 2569.

$$\begin{array}{r} 4534 \\ -1965 \\ \hline 2569 \end{array} \qquad \begin{array}{r} {\scriptstyle 3\ 14\ 12\ 14} \\ \not{4}\ \not{5}\ \not{3}\ \not{4} \\ -1\ 9\ 6\ 5 \\ \hline 2\ 5\ 6\ 9 \end{array}$$

Graphic display of the exchange method

Multiplication

Definition: Multiplication is simplified repeated addition. For example, there are 4 rows of apples. Each row contains 8 apples. The rows can be added $8 + 8 + 8 + 8 = 32$ or more simply, you can multiply $4 \times 8 = 32$.

The terms of multiplication are the multiplicand, which is the number to be multiplied by another number or the number to be repeated; the multiplier, the number which is to multiply another number or the number of times the multiplicand is to be repeated, and the product which is the result of the multiplication. The multiplicand and the multiplier are also known as the factors of the product. A **factor** being each of the numbers or letters multiplied together to obtain a given product. Thus, 3 and 5 are factors of 15.

$$\begin{array}{r} 5 \\ \times 3 \\ \hline 15 \end{array}, \quad \begin{array}{c} \text{multiplicand} \\ \text{multiplier} \\ \hline \text{product} \end{array}$$

Multiplication can be represented in several ways as shown below:

$$5 \times 3 = 15,\ 5 \cdot 3 = 15,\ (5)(3) = 15,\ 5(3) = 15,\ (5)3 = 15$$

Algorithm: Multiplication of single digits is a matter of memorization. When multiplying two signed numbers, if the signs are different (one positive and one negative) the product is negative. If the signs are the same (both positive or both negative) the product is positive. Multiplying an odd number of negative numbers will produce a negative answer. Multiplying an even number of negative numbers will produce a positive number.

Numbers of two or more digits can be multiplied using the carrying method. With the carrying method you multiply the digits from right to left. If the number is greater than the units place value (greater than nine), carry the value of the number greater than the units' value to the next place value and add to the product.

Example 8: Multiplication of positive and negative integers.

$$5 \times 2 = 10$$
$$5 \times (-2) = -10$$
$$-5 \times 2 = -10$$
$$-5 \times (-2) = 10$$

Example 9: Horizontal multiplication with an odd number of negative numbers.

$(-3)(+8)(-1)(-2) = -48$ Three numbers are negative

Example 10: Horizontal multiplication with an even number of negative numbers.

(-4)(+5)(-2)(+3)(+1) = 120 Two numbers are negative

Example 11: Vertical multiplication when the multiplier is a single digit.

$$\begin{array}{r} 3279 \\ \times \quad 8 \\ \hline 26,232 \end{array}$$

Using the carrying method; multiply 8×9; the product equals 72. Place the 2 in the units place of the product and carry the 7 to the tens column. Multiply 8×7 the product equals 56. Add the 7 that was carried over from the unit's column $(56+7=63)$. Place the 3 in the tens column of the product and carry the 6 to the hundreds column. Multiply 8×2 and add the 6 that was carried over from the tens column $(16+6=22)$. Finally, multiply 8×3 and add the 2 that was carried over from the hundreds column $(24+2=26)$. The final answer is 26, 232.

Example 12: Multiplication when the multiplier has two or more digits.

$$\begin{array}{r} 3487 \\ \times \quad 28 \\ \hline 27896 \\ 69740 \\ \hline 97,636 \end{array}$$

Proceed as above for the units' digit of the multiplier (8 in this example). Then multiply by the tens digit of the multiplier (2 in this example). Since you are starting in the tens place of the multiplier, place a zero in the units position of the second partial product. Place the second partial product (6974 in this example) under the tens digit of the first partial product (27896 in this example) and add or sum the two partial products, which gives the final answer of 97,636. In this example, you are multiplying $3487\times8=27896$ and $3487\times2=69740$ and summing the two partial products.

Example 13: Multiplication when there is a zero in the multiplicand.

$$\begin{array}{r} 4208 \\ \times \quad 47 \\ \hline 29456 \\ 168320 \\ \hline 197,776 \end{array}$$

Proceed as above. Since $7 \times 8 = 56$ you have to bring down the 6 and carry the 5 ($0 + 5 = 5$) as shown in the first partial product. The process is the same for the tens digit of the multiplier where $4 \times 8 = 32$ so you bring down the two in the second partial sum and carry the $3(0 + 3 = 3)$

Example 14: Multiplication when there is a zero in the tens and hundreds position of the multiplicand.

$$\begin{array}{r} 4200 \\ \times \quad 47 \\ \hline 29400 \\ 168000 \\ \hline 197,400 \end{array}$$

In this example the process is the same. Since $7 \times 0 = 0$ you bring down the 0 and place it in the units' position of the first partial product. Next, multiply the 0 in the tens position by 7. Again $7 \times 0 = 0$ so you bring down the 0 and place it in the tens position of the first partial product. Proceed as above for the 2 and 4. Since the next partial product starts with a number in the tens position you start by placing a 0 in the units' position of the second partial product.

Example 15: Multiplication when there is a zero in the multiplier.

a.
$$\begin{array}{r} 3246 \\ \times 305 \\ \hline 16230 \\ 973800 \\ \hline 990,030 \end{array}$$

b.
$$\begin{array}{r} 3246 \\ \times \quad 300 \\ \hline 973,800 \end{array}$$

In example 15a, the 0 in 305 is in the tens place so a 0 is placed in the tens position of the second partial product. Finally, the multiplicand is multiplied by 3 to complete the second partial product. In example 15b, the multiplier 300 has a zero in both the units and tens position so two zeros are placed in the product before the multiplicand 3246 is multiplied by 3.

Division

Definition: Division is the process of determining the number of times a given number contains another number. Division is the inverse of multiplication or repeated subtraction. The terms of division are the dividend, or the number to be divided or separated into equal parts; the divisor or the number of equal parts into which the dividend is to be separated or the number by which the dividend is to be divided; and the quotient, which is the result obtained by division. If the divisor is not an even factor of the dividend, the quotient will contain a remainder. The remainder is the integer amount that is left over after division. The slash (/) or obelus (÷) signs are used to indicate division. The unnamed symbol $\overline{)}$ is used to indicate long division.

$$\begin{array}{rl} 15 & \text{dividend} \\ \div\ 3 & \text{divisor} \\ \hline 5 & \text{quotient} \end{array}, \qquad \text{Long division form:} \quad \begin{array}{r} \text{quotient} \\ \text{divisor}\overline{)\text{dividend}} \end{array}, \quad \begin{array}{r} 5 \\ 3\overline{)15} \end{array}$$

Algorithm: Division of single digits is a matter of memorization. When dividing two signed numbers, if the signs of the dividend and the divisor are different (one positive and one negative) the quotient is negative. If the signs of the dividend and the divisor are the same (both positive or both negative) the quotient is positive. The basic procedure for long division is as follows:

1. When dividing two numbers, for example, *n* divided by *m*, *n* is the dividend and *m* is the divisor; the answer is the quotient.
2. Find the location of all decimal points in the dividend and divisor.
3. If necessary, simplify the long division problem by moving the decimals of the divisor and dividend by the same number of decimal places, to the right, (or to the left) so that the decimal of the divisor is to the right of the last digit.
4. When doing long division, keep the numbers lined up straight from top to bottom under the correct place value.
5. After each step, be sure the remainder for that step is less than the divisor. If it is not, there are three possible problems: the multiplication is wrong, the subtraction is wrong, or a greater quotient is needed.
6. In the end, the remainder, *r*, is added to the quotient as a fraction, *r*/*m*.

Example 16: The divisor is a single digit and the dividend has more than two digits. Divide 656 by 4.

$$\begin{array}{r} 164 \\ 4\overline{)656} \\ 4 \\ \hline 25 \\ 24 \\ \hline 16 \end{array}$$

Begin at the left or the first number of the dividend. 4 is contained in 6 one time with a remainder of $2\,(4\times1=4, 6-4=2)$. Place the 1 above the 6 in the quotient. Place a 4 under the 6 in the dividend and subtract. The remainder represents 2 hundreds or 20 tens and must be added to the 5 tens (bring the 5 down) which equals the partial dividend 25. Next, 25 contains 4 six times with a remainder of 1 $(4\times6=24, 25-24=1)$. Place the 6 above the 5 in the quotient. Place a 1 under the 24 in the dividend and subtract. Bring down the 6 which is the last digit of the dividend. Sixteen contains 4, four times with no remainder. The final answer is 164. Check the result by multiplying the divisor (4) times the quotient (164), which gives you the original dividend of 656.

Example 17: The divisor is two or more digits and the dividend has more than two digits. Divide 73,158 by 534.

$$\begin{array}{rl} 137 & \\ 534\overline{)73,158} & \\ \underline{53\,4} & (534\times1) \\ 1975 & \text{(partial dividend)} \\ \underline{1602} & (534\times3) \\ 3738 & \text{(partial dividend)} \\ \underline{3738} & (534\times7) \\ 0 & \end{array}$$

Since the divisor has 3 digits, take the first 3 digits of the dividend 731, and determine how many times the divisor 534 is contained in 731. Start by using the first digit of the divisor as a trial and determining how many times it is contained in the first digit of the dividend (in this case 5 is contained in 7 one time). If the first digit of the divisor is too large to be contained in the first digit of the dividend try dividing the first two digits of the dividend, and so on. Place the partial quotient 1 over the last digit of 731. Subtract $(534\times1=534)$ from 731, which equals 197 and bring down the 5, which working from right to left is the next digit in the dividend. The result is the partial dividend 1975. Divide the first two digits of the partial dividend 19 by 5. Write the partial quotient of 3 over the 5 in the dividend. Subtract $(534 \times 3 = 1602)$ from 1975, which is 373. Bring down the 8, which is the next digit of the dividend. The result is the partial dividend 3738. Divide the partial dividend 3738 by 534. Place the partial quotient 7 over the 8 in the dividend. Subtract $(534 \times 7 = 3738)$ from the partial dividend 3738, which results in 0. The final answer is 137 with no remainder.

Example 18: The partial dividend is less than the divisor and the result contains a remainder. Divide 89,151 by 29.

$$\begin{array}{r} 3074 \\ 29\overline{)89151}, \ 3074\,R5 \\ \underline{87} \\ 215 \\ \underline{203} \\ 121 \\ \underline{116} \\ 5 \end{array} \qquad \textbf{Check:} \quad \begin{array}{r} 3074 \\ \times \quad 29 \\ \hline 27666 \\ 6148 \\ \hline 89146 \\ +5 \\ \hline 89151 \end{array}$$

The divisor 29 has two digits, take the first two digits of the dividend 89. Twenty nine is contained in 89 three times $(3 \times 29 = 87)$. Place the 3 in the quotient above the units' place of 89 and subtract 87 from 89. The result is a remainder of 2. When you bring down the 1 you get 21 which is less than the divisor 29, therefore you place a 0 after the 3 in the quotient and bring down the 5 resulting in the partial dividend of 215. Twenty nine is contained in 215 seven times $(7 \times 29 = 203)$ with a remainder of 12. Bring down the 1, which results in 121. Twenty nine is contained in 121 four times $(4 \times 29 = 116)$ with a remainder of 5. Because 5 will not contain 29 you have a remainder of 5; dividing 5 by 29 gives you a decimal fraction .1724.

The remainder can be written with the answer as 3074 R5 or as $3074\frac{5}{29}$.

Summary

Arithmetic is defined as the study of quantity, especially as the result of combining numbers and is formally defined as the science of number and computation. The operations of arithmetic include addition, subtraction, multiplication and division. Addition is the process of finding the number that is equal to two or more numbers grouped together. Subtraction is the process of finding a quantity which when added to one of two given quantities will give the other. Subtraction is the inverse (opposite in order) of addition. Multiplication is simplified repeated addition and division is the process of determining the number of times a given number contains another number. Division is the inverse of multiplication or repeated subtraction.

Fractions

Definition: A **fraction** (from the Latin *fractus*, broken) is a number that can represent part of a whole (part-to-whole ratio); the quotient of two rational numbers; a part of a unit or an indicated quotient of one number divided by another; or a numerical representation indicating the quotient of two numbers. A fraction is made up of two parts, the numerator and denominator separated by a line. If the line is slanting, it is called a solidus or forward slash. If the line is horizontal, it is called a vinculum. The **numerator** is the dividend of a fraction; it is the number of parts you have. The **denominator** is the divisor of a fraction; it is the number of parts the whole is divided into.

For example, if the fraction is $\frac{3}{5}$ of a pie, the denominator 5 means that the pie has been divided into 5 equal parts, of which there are 3 parts, or 3 out of 5. The operations that apply to fractions are the basic arithmetic operations of addition, subtraction, multiplication, and division, and the process of simplification or reduction.

Fractional From $\dfrac{\text{numerator}}{\text{denominator}}, \dfrac{\text{dividend}}{\text{divisor}}$

There are several types of fractions including **proper fractions**, **improper fractions**, **mixed fractions** and **complex fractions**. The simplest form is the proper fraction.

A **proper fraction** is a fraction in which the numerator is less than the denominator. A proper fraction represents a part of a whole and therefore, has a value less than one.

Since $\frac{4}{4}=1$, decreasing the numerator will result in a value less than one, as shown by the fraction below which has decimal value .75. Decimals will be examined in greater detail following the section on fractions.

$\frac{3}{4} = .75$

An **improper fraction** is a fraction in which the numerator equals or is greater than the denominator or has a value greater than one. For example $\frac{9}{8} = 1\frac{1}{8} = 1.125$, which is greater than one.

A **mixed fraction**, sometimes referred to as a mixed number or mixed expression, is a whole number and a fraction taken together. For example, $3\frac{2}{5}$.

A complex fraction is a fraction that contains one or more fractions in its numerator, in its denominator, or both. For example, $\frac{\frac{1}{2}}{\frac{3}{4}}$.

History: Fractions were used as early as 2800 BC in the ancient Indus Valley civilization. The Indus Valley civilization was a Bronze Age civilization which was centered mostly in the western part of the Indian Subcontinent. The Egyptians used Egyptian fractions ca. 1000 BC. In the Egyptian number system, all fractions except 2/3 had the number 1 as the numerator. Any other fraction was represented as the sum of unit fractions. The Greeks used unit fractions and later continued fractions. The followers of the Greek philosopher Pythagoras (ca. 530 BC) discovered, to their dismay, that the square root of two cannot be expressed as a fraction. This is known today as an irrational number. Al-Hassar, a Muslim mathematician from the Maghreb (North Africa) developed the modern symbolic mathematical notation for fractions, where the numerator and denominator are separated by a horizontal bar. This same fractional notation appears soon after in the work of Fibonacci in the 13th century.

Applications: Fractions are very important in higher level mathematics and are used in the form of rational equations in algebra and fractional exponents in calculus. They are also used in work related fields and in the home. Many jobs make extensive use of fractions including construction, machining, and manufacturing, where fractions are used for measurement and mixtures. In the home, fractions are used in cooking and mixing.

Rules for Fractions

If $\frac{a}{b}$ and $\frac{c}{d}$ are fractions with $b \neq 0$ and $d \neq 0$ then:

1. Equality: $\frac{a}{b} = \frac{c}{d} \Leftrightarrow ad = bc$

Example: $\frac{1}{2} = \frac{4}{8} = (1)(8) = (2)(4)$

2. Equivalency: $\frac{a}{b} \cdot \frac{c}{c} = \frac{ac}{bc}$

Example: $\frac{2}{3} \cdot \frac{2}{2} = \frac{4}{6}$ or in reduced form $\frac{2}{3}$

3. Addition and subtraction (like denominators): $\frac{a}{b} \pm \frac{c}{b} = \frac{a \pm c}{b}$

Example: $\frac{2}{3} + \frac{3}{3} = \frac{5}{3}$

4. Addition and subtraction (unlike denominators):

$$\frac{a}{b} \pm \frac{c}{d} = \frac{ad}{bd} \pm \frac{cb}{bd} = \frac{ad \pm cb}{bd}$$

Example: $\frac{1}{2} + \frac{2}{3} = \frac{(1)(3)}{(2)(3)} + \frac{(2)(2)}{(2)(3)} = \frac{3+4}{6} = \frac{7}{6}$

5. Multiplication: $\frac{a}{b} \cdot \frac{c}{d} = \frac{ac}{bd}$

Example: $\frac{1}{2} \cdot \frac{2}{3} = \frac{2}{6} = \frac{1}{3}$

6. Division (invert and multiply): $\frac{a}{b} \div \frac{c}{d} = \frac{a}{b} \cdot \frac{d}{c} = \frac{ad}{bc}$

Example: $\frac{1}{2} \div \frac{2}{3} = \frac{1}{2} \cdot \frac{3}{2} = \frac{3}{4}$

$$\frac{a}{\left(\frac{b}{c}\right)} = \frac{\left(\frac{a}{1}\right)}{\left(\frac{b}{c}\right)} = \left(\frac{a}{1}\right)\left(\frac{c}{b}\right) = \frac{ac}{b}$$

$$\frac{\left(\frac{a}{b}\right)}{c} = \frac{\left(\frac{a}{b}\right)}{\left(\frac{c}{1}\right)} = \left(\frac{a}{b}\right)\left(\frac{1}{c}\right) = \frac{a}{bc}$$

7. Complex fractions:

$$\frac{\left(\frac{a}{b}\right)+\left(\frac{c}{d}\right)}{\left(\frac{e}{f}\right)+\left(\frac{g}{h}\right)}=\frac{\left(\frac{a}{b}\right)+\left(\frac{c}{d}\right)}{\left(\frac{e}{f}\right)+\left(\frac{g}{h}\right)}\cdot\frac{bdfh}{bdfh}=\frac{(ad+bc)fh}{(eh+fg)bd}$$

Example: $$\frac{\left(\frac{1}{2}\right)+\left(\frac{2}{3}\right)}{\left(\frac{3}{4}\right)+\left(\frac{2}{5}\right)}=\frac{[(1)(3)+(2)(2)](4)(5)}{[(3)(5)+(4)(2)](2)(3)}=\frac{(3+4)20}{(15+8)6}=\frac{140}{138}=\frac{70}{69}$$

Common Fraction Errors

$\frac{a}{b+c}\neq\frac{a}{b}+\frac{a}{c}$ But $\frac{b+c}{a}=\frac{b}{a}+\frac{c}{a}$

$\frac{a+bx}{a}\neq 1+bx$ The a in the numerator is not a factor

$\frac{a}{0}\neq 0$ Division by zero is undefined

Tools for Fractions

Adding and subtracting fractions requires that the fractions have a common denominator. To simplify these operations it is helpful to find the lowest common denominator (LCD). Answers to problems involving fractions are generally required to be reduced to lowest terms. This is done by dividing both the numerator and denominator by the largest number that will divide evenly into both. This number or expression is known as the greatest common divisor (GCD).

Least common multiple (LCM) – the smallest number that is divisible by each member of a set of numbers; a multiple is a quantity containing another quantity a number of times without a remainder. For example, for the set containing 10 and 4 the LCM is 20; to find the LCM of two or more expressions, find the prime factors of each expression; take each factor the greatest number of times it occurs in any one of the expressions and find the product of the different prime factors. Prime factors are factors that are prime numbers. As stated earlier, prime numbers are numbers that can only be divided by 1 and the number itself.

Example 1: Find the LCM of 12 and 30.

2|12	2|30	$12 = 2 \times 2 \times 3$
2|6	3|15	$30 = 2 \times 3 \times 5$
3|3	5|5	$LCM = 2 \times 2 \times 3 \times 5 = 60$

Begin by dividing with the lowest prime number that is a factor of the numbers for which you are trying to determine the LCM, which in this case is 2. Keep dividing by 2 until the remainder is no longer evenly divisible by two. Then use the next lowest prime factor, which is 3. Repeat the process until the number is only divisible by 1 or itself. Group the prime factors which occur the largest number of times in any of the given numbers and multiply as shown above. In this example, 2 occurs twice in the factorization of 12, while 3 and 5 only occur once in the factorization of both 12 and 30. Multiply these prime factors, which results in the LCM being equal to 60.

Example 2: Find the least common multiple (LCM) of 180, 360, and 450.

2|180	2|360	2|450
2|90	2|180	3|225
3|45	2|90	3|75
3|15	3|45	5|25
5|5	3|15	5|5
	5|5	

Begin by dividing 180, 360 and 450 by the lowest or smallest prime factor, which is 2. Continue with next prime factor that will evenly divide the remainder. The final answer is 1800.

$$2 \times 2 \times 2 \times 3 \times 3 \times 5 \times 5 = 1800$$

Least Common Denominator (LCD) - or lowest common denominator is the **least common multiple (LCM)** of the denominators of a set of fractions. It is the smallest positive integer that is a multiple of the denominators. A common multiple is a number which is a multiple of two or more integers. You can always get a common multiple of two integers by multiplying them, but unless the two numbers are relative primes, the product will not be the least common multiple. The "cross-multiply" method of comparing fractions effectively creates a common denominator by multiplying both denominators together but the denominator can become large very quickly.

Example 3: Find the LCD of $\frac{5}{12}$ and $\frac{11}{18}$.

$2\underline{|12}$ $\quad$ $2\underline{|18}$

$2\underline{|6}$ $\quad$ $3\underline{|9}$

$3\underline{|3}$ $\quad$ $3\underline{|3}$ $\quad$ $2\times2\times3\times3=36$ $\quad$ LCD = 36

Follow the same process as above to find the LCM. To convert the original fractions to fractions with a common denominator, determine the number of times 12 is contained in 36, which is 3, and multiply $\frac{5}{12}\times\frac{3}{3}$, ($\frac{3}{3}$ is equal to 1). Then determine how many times 18 is contained in 36, which is 2 and multiply $\frac{11}{18}\times\frac{2}{2}$, (again $\frac{2}{2}$ is equal to 1). The result is shown below.

$$\frac{5}{12}\times\frac{3}{3}=\frac{15}{36} \qquad \frac{11}{18}\times\frac{2}{2}=\frac{22}{36}$$

Finding LCD's and using them to add, subtract, reduce, and compare magnitudes of fractions will be examined in greater detail below.

Greatest Common Divisor (GCD) – The GCD is also referred to as the greatest common factor. The GCD is the greatest number or expression that is a factor of two or more numbers or expressions. There are three methods for finding the GCD, the factoring method, the shortcut method, and Euclid's method named after the Greek mathematician. The factoring method is based on the overlap (have something in common) in the prime factorizations of the numbers under consideration. The same method as used in finding the LCM can be used to find the prime factors for the GCD.

Example 4: Find the GCD of 84 and 96 using the factoring method.

$2\underline{|84}$

$2\underline{|42}$

$3\underline{|21}$

$7\underline{|7}$

$84=2\times2\times3\times7$

$2\underline{|96}$

$2\underline{|48}$

$2\underline{|24}$

$2\underline{|12}$

$2\underline{|6}$

$3\underline{|3}$

$96=2\times2\times2\times2\times2\times3$

Begin by dividing each number by the smallest prime factor which in this case is 2. Keep dividing by the next highest prime number until all prime factors have been found. The numbers that overlap (are contained in both numbers) are $2 \times 2 \times 3 = 12$ so the greatest common factor is 12.

Example 5: Find the GCD of 42, 60, and 80 using the shortcut method.

$$\begin{array}{r|l} 2 & 42,60,80 \\ \hline 3 & 21,30,42 \\ \hline & 7,10,14 \\ \hline \end{array} \qquad 2 \times 3 = 6$$

Arrange the numbers as shown above and begin dividing by the smallest prime number greater than one that will evenly divide all three numbers. In this example, two is the smallest prime number. Continue this process until no divisor can be found that will evenly divide the last quotient. Then multiply the common factors to find the GCD.

Euclid's method for finding the GCD of two numbers is by a process of repeated division. The larger of the two numbers is divided by the smaller, and the new remainder is divided into the previous remainder until a remainder of zero is obtained.

Example 6: Find the GCD of 4,284 and 14,586 using Euclid's method.

$$\begin{array}{r} 3 \\ 4284\overline{)14586} \\ 12852 \\ \hline 1734 \end{array} \qquad \begin{array}{r} 2 \\ 1734\overline{)4284} \\ 3468 \\ \hline 816 \end{array} \qquad \begin{array}{r} 2 \\ 816\overline{)1734} \\ 1632 \\ \hline 102 \end{array} \qquad \begin{array}{r} 8 \\ 102\overline{)816} \\ 816 \\ \hline 0 \end{array}$$

Euclid's method for finding the GCD of two numbers uses a process of repeated division. The larger of the two numbers is divided by the smaller number. Then the remainder is divided into the smaller number and the new remainder divided into the previous remainder. This process is continued until a remainder of 0 is obtained. The last non-zero remainder is the GCD. In this example, the last non-zero remainder is 102, which is the GCD.

Simplification of Fractions

Simplification of a fraction or reducing a fraction to lowest terms is a fraction in which all common factors have been divided out of the numerator and denominator. To reduce a fraction to its lowest terms obtain an equivalent fraction by dividing the numerator and denominator by the largest common factor or the greatest common divisor.

Example 7: Reduce $\frac{18}{27}$ to its lowest terms.

$$\begin{array}{r|l} 2 & 18 \\ 3 & 9 \\ 3 & 3 \\ & 1 \end{array} \qquad \begin{array}{r|l} 3 & 27 \\ 3 & 9 \\ 3 & 3 \\ & 1 \end{array}$$

The numbers that overlap are $3 \times 3 = 9$ so 9 is the GCD

Divide the numerator and the denominator by 9 which equals $\frac{2}{3}$. The fraction $\frac{2}{3}$ cannot be reduced further which means that $\frac{18}{27}$ has now been reduced to its lowest terms. You could also list the factors of 18 (2, 3, 6, and 9) and the factors of 27 (3 and 9) and see that the largest common factor of 18 and 27 is 9.

Simplification of fractions can also be used as a short cut in the process of multiplication and division of fractions. If all common factors to both the numerator and denominator are divided out or canceled (see cancelation below) the result of the multiplication or division will already be in lowest terms.

Example 8:

$$\frac{2}{4} \times \frac{3}{6} \times \frac{4}{3} = \frac{\cancel{2}}{2 \times \cancel{2}} \times \frac{\cancel{3}}{\cancel{2} \times 3} \times \frac{\cancel{2} \times 2}{\cancel{3}} = \frac{2}{6} = \frac{1}{3}$$

Arithmetic Operations Using Fractions

Adding Fractions Algorithm: If the denominators of the fractions are the same, add the numerators and place the resulting sum over the common denominator. If the denominators are not the same, you have to find a common denominator using the least common denominator (LCD) method described earlier, before you can add the numerators. Reduce to lowest terms.

Example 9: Add $\frac{2}{5} + \frac{3}{5} + \frac{4}{5}$.

$$\frac{2}{5} + \frac{3}{5} + \frac{4}{5} = \frac{2+3+4}{5} = \frac{9}{5} = 1\frac{4}{5}$$

Example 10: Add $\frac{1}{2} + \frac{2}{3} + \frac{5}{6}$.

$$\frac{1}{2}+\frac{2}{3}+\frac{5}{6}=\frac{3}{6}+\frac{4}{6}+\frac{5}{6}=\frac{3+4+5}{6}=\frac{12}{6}=2$$

In this example, the denominators are not the same. The lowest common denominator is 6. To convert $\frac{1}{2}$ to a fraction with a denominator of 6 multiply both the numerator and denominator by three $\left(\frac{1}{2}\times\frac{3}{3}\right)$, which is equal to $\frac{3}{6}$. To convert $\frac{2}{3}$ to a fraction with a denominator of 6 multiply both the numerator and denominator by two $\left(\frac{2}{3}\times\frac{2}{2}\right)$, which is equal to $\frac{4}{6}$. The fraction $\frac{5}{6}$ already has a denominator of 6 so no changes are necessary. Multiplying by $\frac{3}{3}$ and $\frac{2}{2}$ is equivalent to multiplying by 1 so the values do not change.

Subtracting Fractions Algorithm: If the denominators of the fractions are the same, subtract the numerators and place the resulting remainder over the common denominator. If the denominators are not the same, you have to find a common denominator using the least common denominator (LCD) method described earlier, before you can subtract the numerators. Reduce to lowest terms.

Example 11: Subtract $\frac{3}{8}$ from $\frac{7}{8}$.

$$\frac{7}{8}-\frac{3}{8}=\frac{7-3}{8}=\frac{4}{8}=\frac{1}{2}$$

The preliminary answer $\frac{4}{8}$ has to be reduced to lowest terms. As stated above to reduce a fraction to lowest terms we use the greatest common divisor (GCD). In this example, the GCD is 4. Dividing both the numerator and denominator by 4 we get the final answer $\frac{1}{2}$.

Example 12: Subtract $\frac{3}{10}$ from $\frac{4}{5}$.

$$\frac{4}{5}-\frac{3}{10}=\frac{8}{10}-\frac{3}{10}=\frac{5}{10}=\frac{1}{2}$$

In this example we use both the LCD to get a common denominator of 10, and the GCD to get a common divisor of 5 to reduce the fraction to lowest terms.

Multiplying Fractions Algorithm: To multiply one proper fraction by another place the product of the numerators over the product of the denominators. To multiply a proper fraction by a whole number either multiply the numerator, or divide denominator by the whole number. To multiply two mixed numbers, change the mixed numbers to improper fractions and multiply as you would with two proper fractions.

Example 13: Multiply the two proper fractions $\frac{3}{4}\times\frac{7}{8}$.

$$\frac{3}{4}\times\frac{7}{8}=\frac{3\times7}{4\times8}=\frac{21}{32}$$

Example 14: Multiply the proper fraction $\frac{7}{16}$ times the whole number 4:

$$\frac{7}{16}\times4=\frac{7\times4}{16}=\frac{28}{16}=\frac{7}{4}$$

Example 15: Multiply the two mixed fractions $2\frac{2}{3}\times3\frac{2}{3}$.

$$2\frac{2}{3}\times3\frac{2}{3}=\frac{(3\times2)+2}{3}\times\frac{(3\times3)+2}{3}=\frac{8}{3}\times\frac{11}{3}=\frac{88}{9}=9\frac{7}{9}$$

In this example, you convert the mixed fractions to improper fractions by multiplying the denominator times the whole number part of the fraction and adding the numerator. Then multiply the improper fractions and simplify.

Dividing Fractions Algorithm: To divide fractions, invert (turn upside down) the fraction that is the divisor and multiply the inverted fraction by the fraction that is the dividend. This is the same as multiplying by the reciprocal. The **reciprocal** of a number is the number whose product with a given number is equal to one. Reciprocal quantities are any two quantities which produce one when multiplied together. For a fraction, the reciprocal is the fraction formed by interchanging the numerator and the denominator in the given fraction. For example, the reciprocal of $\frac{3}{1}$ is $\frac{1}{3}$ since $\frac{3}{1}\times\frac{1}{3}=1$.

Example 16: Divide the proper fractions $\frac{3}{7} \div \frac{3}{14}$.

$\frac{3}{7} \div \frac{3}{14} = \frac{3}{7} \cdot \frac{14}{3} = \frac{42}{21} = 2$

Here $\frac{3}{7}$ (the dividend) is being divided by $\frac{3}{14}$ (the divisor). Multiply the dividend by the reciprocal of the divisor which is $\frac{14}{3}$.

Example 17: Divide the proper fraction $\frac{8}{9}$ by the whole number 2.

$$\frac{8}{9} \div 2 = \frac{\frac{8}{9}}{\frac{2}{1}} = \frac{8}{9} \cdot \frac{1}{2} = \frac{8}{18} = \frac{4}{9}$$

Example 18: Divide the whole number 2 by the fraction $\frac{8}{9}$.

$$\frac{2}{\frac{8}{9}} = \frac{\frac{2}{1}}{\frac{8}{9}} = \frac{2}{1} \cdot \frac{9}{8} = \frac{18}{8} = 2\frac{2}{8} = 2\frac{1}{4}$$

Place the 2 over 1 and invert $\frac{8}{9}$ and multiply

Example 19: Divide the mixed numbers $9\frac{4}{5} \div 2\frac{7}{8}$.

$$9\frac{4}{5} \div 2\frac{7}{8} = \frac{49}{5} \div \frac{23}{8} = \frac{49}{5} \cdot \frac{8}{23} = \frac{392}{115} = 3\frac{47}{115}$$

Change the mixed fractions to improper fractions, invert the divisor, and multiply.

Simplifying Complex Fractions

As defined earlier, a **complex fraction** is a fraction that contains one or more fractions in its numerator, in its denominator, or both. To change a complex fraction to a simple fraction, multiply the numerator by the reciprocal of the denominator, or multiply both numerator and denominator by a common multiple of the denominators of the fractions.

Example 20: Simplify the complex fraction $\frac{\frac{2}{3}}{\frac{4}{7}}$ using the reciprocal method.

$$\frac{\frac{2}{3}}{\frac{4}{7}} = \frac{2}{3} \times \frac{7}{4} = \frac{14}{12} = \frac{7}{6} = 1\frac{1}{6}$$

Example 21: Simplify the complex fraction $\frac{\frac{2}{3}}{\frac{4}{7}}$ using the LCD method.

$$\frac{\frac{2}{3}}{\frac{4}{7}} = \frac{\frac{2}{3} \cdot \frac{21}{1}}{\frac{4}{7} \cdot \frac{21}{1}} = \frac{\frac{42}{3}}{\frac{84}{7}} = \frac{14}{12} = \frac{7}{6} = 1\frac{1}{6}$$

Negative Fractions

Due to the fact that fractions are a representation of division, when you divide a positive number by a negative number you get a negative number. Therefore, the following fractions are all equivalent:

$$-\frac{3}{4} = \frac{-3}{4} = \frac{3}{-4}$$

Magnitude of Fractions

To compare the magnitude or the relative size of two or more fractions, multiply each fraction so that they all have a common denominator.

Example 22: Which of the following two fractions is the largest, $\frac{5}{8}$ or $\frac{3}{32}$?

$$\frac{5}{8} \times \frac{4}{4} = \frac{20}{32}$$

Now it can be seen that $\frac{5}{8}$ or $\frac{20}{32}$ is greater than $\frac{3}{32}$.

Behavior of Fractions

The behavior of fractions is important when determining limits in calculus. The four most interesting behaviors are as follows:

1. If the denominator remains constant and the numerator gets larger the value of the fraction gets larger. Conversely, if the denominator remains constant and the numerator gets smaller the value of the fraction gets smaller. For example,

   $$\frac{1}{4}=.25,\ \frac{2}{4}=.50,\ \frac{3}{4}=.75,\frac{4}{4}=1.0,\ \frac{5}{4}=1.25.$$

2. If the numerator remains constant and the denominator gets larger the value of the fraction gets smaller. Conversely, if the numerator remains constant and the denominator gets smaller the value of the fraction gets larger. For example,

   $$\frac{3}{.01}=300,\ \frac{3}{1}=3.0,\ \frac{3}{2}=1.5,\ \frac{3}{3}=1.0,\ \frac{3}{4}=.75$$

Cancelation

Cancelation is the act of dividing like factors out of the numerator and denominator of a fraction. In the example below, the numerator and denominator were first factored then the like factors were divided out or canceled. In this case, there were two 2's and one 3 that were like factors in both the numerator and denominator. A number divided by itself is one, so you are essentially multiplying the fraction by one.

$$\frac{2}{4}\times\frac{3}{6}\times\frac{4}{3}=\frac{\cancel{2}}{2\times\cancel{2}}\times\frac{\cancel{3}}{\cancel{2}\times 3}\times\frac{\cancel{2}\times 2}{\cancel{3}}=\frac{2}{6}=\frac{1}{3}$$

Summary

A fraction is a number that can represent part of a whole (part-to-whole ratio); the quotient of two rational numbers; a part of a unit or an indicated quotient of one number divided by another; or a numerical representation indicating the quotient of two numbers. There are several types of fractions including proper fractions, improper fractions, mixed fractions and complex fractions.

Fractions are difficult for many people because there are multiple definitions and types of fractions, and a fairly large number of rules and computations that are required to solve fraction problems. The operations of arithmetic including addition, subtraction, multiplication and division can all be applied to fractions. Additional computations required to solve fraction problems include Least Common Multiple (LCM), Least Common Denominator (LCD), and Greatest Common Denominator (GCD). The LCM is the smallest number that is divisible by each member of a set of numbers. The LCD is the least common multiple of the denominators of a set of fractions. It is the smallest positive integer that is a multiple of the denominators. The GCD is the greatest number or expression that is a factor of two or more numbers or expressions.

Due to the fact that fractions are a representation of division, when you divide a positive number by a negative number you get a negative number. Therefore, the placement of the negative sign does not change the sign of the fraction. To determine the comparative magnitude of fractions multiply each fraction so that they all have a common denominator. If the denominator remains constant and the numerator gets larger the value of the fraction gets larger. If the numerator remains constant and the denominator gets larger the value of the fraction gets smaller.

Decimals

Definition: A **decimal** is any proper fraction in which the denominator is some power of ten. The denominators are not usually written but are indicated by the use of a dot, called the decimal point. The **decimal point** is a dot or full stop at the left of a decimal fraction. The figures at the left of the point represent units or whole numbers, such as 1.05. The figures to the right of the decimal point represent the fractional part of the decimal. For example:

$$\frac{1}{10} = .10 \qquad \frac{1}{5} = \frac{2}{10} = .2$$

$$\frac{1}{100} = .01 \qquad \frac{1}{4} = \frac{25}{100} = .25$$

$$\frac{1}{1000} = .001 \qquad \frac{3}{8} = \frac{375}{1000} = .375$$

Algorithm: To add decimal fractions:

1. Write the addends so that the decimal points are aligned in a column.
2. Add zeros to decimals, if necessary, so that you have the same number of decimal places in each addend.
3. Sum the addends
4. Place the decimal point in the sum so that it is aligned with the decimal point in the addends.

Example 1: Add the following numbers $236.146+42.8+125.46+1284.9$.

$$\begin{array}{r} 236.146 \\ 42.800 \\ 125.460 \\ 1284.900 \\ \hline 1689.306 \end{array}$$

Algorithm: To subtract decimal fractions:

1. Write the subtrahend and minuend so that the decimal points are aligned in a column.
2. Add zeros to decimals, if necessary, so that you have the same number of decimal places in both the subtrahend and the minuend.
3. Subtract using the same process for subtraction as stated above.
4. Place the decimal point in the difference or remainder so that it is aligned with the decimal point in the subtrahend and the minuend.

Example 2: Find the difference or remainder of 156.5 and 38.875.

$$\begin{array}{r} 156.500 \\ -38.875 \\ \hline 117.625 \end{array}$$

Algorithm: To multiply decimal fractions:

1. Multiply using the same process for multiplication as stated above.
2. Place a decimal point in the product, so that it will have the same number of decimal places as the sum of the decimal places in the numbers being multiplied.

Example 3: Find the product of 45.87 and 3.25.

$$\begin{array}{r} 45.87 \\ \times 3.25 \\ \hline 22935 \\ 9174 \\ 13761 \\ \hline 149.0775 \end{array}$$

In this example, the sum of the decimals in the numbers being multiplied is 4 so you count 4 places to the left to place the decimal point. If the sum of the decimals in the numbers being multiplied is greater than the values in the product, you place a zero in front of the number for each decimal place and then place the decimal.

Example 4: Find the product of .0005 and .005.

$$\begin{array}{r} .0005 \\ \times .005 \\ \hline 0025 \\ 00000 \\ 000000 \\ \hline .0000025 \end{array}$$

In this example, there were six place values but seven decimal places so you place a zero in front of sixth decimal place and place the decimal point.

Algorithm: To divide decimal fractions:

1. If the divisor is a decimal fraction, change it to a whole number by moving its decimal point the right as many places as necessary.
2. Move the decimal point of the dividend the same number of places to the right as the divisor, adding zeros if necessary.
3. Divide using the same process for division as stated above.
4. Place a decimal point in the quotient equal to the decimal places as changed in the dividend.

Example 1: Find the quotient of 34.944 divided by .24.

$$\begin{array}{r} 145.6 \\ 24\overline{)3494.4} \\ 24 \\ \hline 109 \\ 96 \\ \hline 134 \\ 120 \\ \hline 144 \\ 144 \\ \hline 0 \end{array}$$

In this example, since you move the divisor's decimal point two places to the right, you need to move the dividends decimal point two places to the right.

To **convert a common fraction to a decimal**, place a decimal point after the numerator and add as many zeros as required until the denominator will evenly divide the numerator. To **convert a decimal to a common fraction** express the decimal as a numerator over a denominator and reduce to lowest terms.

Example 1: Convert $\frac{3}{4}$ to a decimal.

$$\frac{3}{4} = \frac{3.00}{4} = .75$$

In this case, 4 will not go into 3 so you have to add a decimal point and a zero after the 3. You then have a remainder of 2 so you have to add another zero.

Example 2: Convert $\frac{5}{9}$ into a decimal.

$$\frac{5}{9} = \frac{5.0000}{9} = .5555...$$

Example 3: Convert 0.125 to a fraction.

$$.125 = \frac{125}{1000} = \frac{1}{8}$$

In this case you have 3 decimal places, which is one thousandth, so you place 125 over 1000 and reduce the fraction to lowest terms. The GCD is 125.

Example 3: Convert 0.333 to a fraction.

$$.333 = \frac{333}{1000} = \frac{37}{111}$$

The GCD is 9

Summary

A decimal is any proper fraction in which the denominator is some power of ten. The denominators are not usually written but are indicated by the use of a dot, called the decimal point. The decimal point is a dot or full stop at the left of a decimal fraction. The operations of arithmetic including addition, subtraction, multiplication and division can all be applied to decimal fractions. Common fractions can be converted to a decimal, and decimals can be converted to a common fraction.

Exponents, Logarithms, and Radicals

Exponents

Definition: An **exponent** is a number placed at the right of and above a number or a symbol. The value assigned to the symbol with this exponent is called a power and indicates the power taken or how many times the number or symbol is multiplied by itself. The terms power and exponent are used interchangeably. The operation of raising quantities or terms to a given power is called involution. The number to be raised to a power is called the base. For example, x^3 pronounced "*x* cubed, or *x* to the third power", equals $x \times x \times x$. If *x* is a nonzero number, the value x^0 is defined to be one. For example, $5^0 = 1$. If $x \neq 0$, x^0 can be thought of as the result of subtracting exponents when dividing a quantity by itself. For example, $\frac{x^2}{x^2} = x^0 = 1$. A negative exponent indicates that in addition to the operations indicated by the numerical value of the exponent, the quantity is to be reciprocated. For example, $x^{-2} = \frac{1}{x^2}$, $3^{-2} = (9)^{-1} = \frac{1}{9}$.

For any number *a* and any counting number *n*, $a^n = a \times a \times a ... \times a$. The number *a* is the base, *n* is the exponent, and a^n is read "*a* to the power *n*, or *a* to the *n*th power."

Fractional Exponents

A **fractional exponent** is an exponent expressed as fraction that indicates a root of an expression. The numerator indicates the power to which the base is to be raised, and the denominator, the root which is to be extracted of that power. The rules of proper fractions apply to fractional exponents.

Example 1: What is the fractional exponent of $\sqrt[3]{x^4}$?

$x^{\frac{4}{3}}$ This is read "the cube root of x to the fourth power."

A **negative fractional exponent** indicates the reciprocal of the expression.

Example 2: The negative fractional exponent $x^{-\frac{1}{3}}$ is equivalent to $\frac{1}{x^3}$.

Operations with Exponents

To multiply two numbers with exponents, if the base numbers are the same, keep the base number and add the exponents. For example, $2^3 \times 2^5 = 2^8$.

To divide two numbers with exponents, if the base numbers are the same, keep the base number and subtract the second exponent from the first. For example, $3^4 \div 3^2 = 3^2$

.

Rules for Exponents

1. Multiplication: $a^n a^m = a^{n+m}$

   Example: $a^2 a^3 = a^5$

2. Division: $\frac{a^n}{a^m} = a^{n-m}$

   Example: $\frac{a^5}{a^3} = a^2$

3. Power rule: $(a^n)^m = a^{nm}$

   Example: $(a^2)^3 = a^6$

4. Distributed over a simple product: $(ab)^n = a^n b^n$

   Example: $(ab)^2 = a^2 b^2$.

5. Distributed over a complex product: $(a^m a^p)^n = a^{mn} b^{pn}$

   Example: $(a^3 a^2)^2 = a^6 b^4$

6. Distributed over a simple quotient: $\left(\frac{a}{b}\right)^n = \frac{a^n}{b^n}$

   Example: $\left(\frac{a}{b}\right)^2 = \frac{a^2}{b^2}$

7. Distributed over a complex quotient: $\left(\frac{a^m}{b^p}\right)^n = \frac{a^{mn}}{b^{pn}}$

Example: $\left(\frac{a^3}{b^4}\right)^2 = \frac{a^6}{b^8}$

8. Negative exponent: $a^{-n} = \frac{1}{a^n}, a^{-1} = \frac{1}{a}$

Example: $a^{-2} = \frac{1}{a^2}$

9. Exponent of zero: $a^0 = 1$

Example: $5^0 = 1$

10. No exponent: $a = a^1$

Example: $5 = 5^1$

11. Fractional exponent: $a^{\frac{1}{n}} = \sqrt[n]{a}$

Example: $a^{\frac{1}{3}} = \sqrt[3]{a}$

12. Negative fractional exponent: $a^{-\frac{m}{n}} = \frac{1}{a^{\frac{m}{n}}}$

Example: $a^{-\frac{2}{3}} = \frac{1}{a^{\frac{2}{3}}} = \frac{1}{\sqrt[3]{a^2}}$

13. Fractional exponent numerator $\neq 1$: $a^{\frac{m}{n}} = \sqrt[n]{a^m}$ or $\left(\sqrt[n]{a}\right)^m$, $(a \geq 0)$

Example: $a^{\frac{2}{3}} = \sqrt[3]{a^2}$

14. Exponential towers (iterated exponentiation): ${}^n a = a^{a^{\cdot^{\cdot^{a}}}}$ an exponential a multiplied by itself n times.

Example: ${}^4 2 = 2^{2^{2^2}} = 2\left[2^{\left(2^2\right)}\right] = 2\left(2^4\right) = 2^{16} = 65,536$

Exponential towers are evaluated from top to bottom.

Logarithms

Definition: A **logarithm** is the exponent, *n*, to which the base *b* must be raised to equal *a*, written as $\log_b a = n$. For example, $\log_2 32 = 5$. This is read "the log of 32 to the base 2 is 5" and means that when the base 2 has an exponent of 5, or 2 is raised to the 5th power, it is equal to 32. Logarithms which use 10 as base are common logarithms. Logarithms which use the base $e = 2.71828\ldots$ are natural logarithms. Any positive number except 1 can act as the base for a logarithm. Since the logarithm of 10 in the base *e* is 2.30259, a common logarithm may be changed to a natural logarithm by multiplying it by 2.30259.

Definition: An **antilogarithm** (abbreviated antilog) is the inverse of the common logarithm. The antilog *x* is simply 10^x. If your calculator has a 10^x key, you obtain the antilog of a number by entering the number and pressing the 10^x key. For example, the antilog of 1.59770 is 39.60. A logarithm is an exponent, therefore $10^{1.59770} = 39.60$.

History: Logarithms were invented by the Scottish mathematician John Napier (1550 – 1617). Napier coined the term logarithm from two Greek words *logos* (ratio) and *arithmos* (number) to describe the theory that he spent twenty years developing and that first appeared in the book *A Description of the Marvelous Rule of Logarithms.* Henry Briggs followed with his table of common logarithms (base ten) of the numbers 1 to 1000 carried out to fourteen places.

Applications: Logarithms and logarithmic functions are used extensively in calculus, including finding antiderivatives, and the simplification of differentiation involving products, quotients and powers. Logarithms are also used in statistics for smoothing graphs of nonlinear functions, for solving complex equations, and in chemistry for the pH scale.

Rules for Logarithms

1. $\ln 1 = 0$, $\log_b 1 = 0$
2. $\ln e = 1$, $\log_b b = 1$
3. $\ln e^x = x$, $\log_b b^x = x$
4. If $b^x = y$ and $b > 0$ then $y = \log_b x$

   Example: $\log_5 125 = 3$ because $5^3 = 125$
5. Logarithm of a base to a power: $\log_b b^x = x$
6. Base to a logarithm: $b^{\log_b x} = x$
7. Notation for logarithm base 10: $\log x = \log_{10} x$
8. Notation for logarithm base *e*: $\ln x = \log_e x$
9. Product rule: $\log_b (MN) = \log_b M + \log_b N$

   Example: $\log_4 5x^3 y = \log_4 5 + \log_4 x^3 + \log_4 y$

10. Quotient rule: $\log_b\left(\frac{M}{N}\right) = \log_b M - \log_b N$

Example: $\ln\left(\frac{\sqrt{3x-5}}{7}\right) = \ln(3x-5)^{\frac{1}{2}} - \ln 7$

11. Power rule: $\log_b N^p = p\log_b N$, $\ln y^x = x\ln y$

Example: $\log_4 x^3 = 3\log_4 x$

12. Change of base formula: $\log_b N = \frac{\log_a N}{\log_a b}$

Example: $\log_e 10 = \frac{\log 10}{\log 2.718} \approx \frac{1}{0.4342} \approx 2.303$

The $\approx$ symbol means approximately.

Example: The acidity of a solution depends on the hydrogen-ion concentration. Acidity is measured using the pH scale which is the negative of the logarithm of the molar hydrogen-ion concentration. The molar concentration M or molarity is the moles (number of molecules) of solute (a gas or solid dissolved in a liquid) dissolved in one liter of solution. A neutral solution has a pH of 7. For an acidic solution the pH is less than 7, and a basic solution has a pH greater than 7. The formula for pH is shown below.

$$\mathrm{pH} = -\log\left[\mathrm{H}^+\right]$$

Where log is the common logarithm or base 10 and H^+ represents the hydrogen ion concentration.

A sample of orange juice has a hydrogen-ion concentration of $2.9\times10^{-4}\,M$. What is the pH? Is the solution acidic?

$$\begin{aligned}\mathrm{pH} &= -\log\left[\mathrm{H}^+\right] \\ &= -\log(2.9\times10^{-4}) \\ &= 3.54\end{aligned}$$

To get the answer with a calculator that supports scientific notation, enter the hydrogen-ion concentration (in exponential notation using the EXP key) and press the LOG key, then change the sign of the result. The pH is less than 7 so the solution is acidic.

Radicals

Definition: A **radical** is the indicated root of a quantity. It is an expression used to indicate the root of a number. It consists of the number of which the root is to be taken, called the **radicand**, the symbol $\sqrt{\ }$ is called the **radical**, and the indicated root *n* ($\sqrt[n]{\ }$) is called the **index**. For example, $\sqrt[3]{64}$ (read the cube root of 64), 64 is the radicand and 3 is the index. The index is generally omitted for square roots. The **root** of a number is always one of the equal factors of that number. Evolution is the inverse of involution (raising to a power). In evolution, the problem is to determine one of a given number of equal factors when their product alone is given. These factors are called roots. An *n*th root of a number is a number which, when taken as a factor *n* times (raised to the *n*th power), produces the given number. There are *n* *n*th roots of any nonzero number. For example, since $9 = 3 \times 3$, then 3 is a root of 9. Using the radical sign, $\sqrt{9} = 3$, or the square root of 9 is three. This follows from the fact that $3^2 = 9$.

Rules for Radicals

A **radical** is an expression used to indicate the root of a number. The components of a radical are as follows:

$\sqrt[n]{a^m}$ Where *n* is the index, $\sqrt{\ }$ is the radical sign, *a* is the radicand, and *m* is the exponent or power of *a*.

If $\sqrt[n]{a}$ and $\sqrt[n]{b}$ are real numbers then:

1. If *n* is even, then: $\sqrt[n]{a^n} = |a|$ Where $|a|$ is the absolute value of *a*

   Example: $\sqrt[2]{5^2} = |5|$

2. In *n* is odd, then: $\sqrt[n]{a^n} = a$

   Example: $\sqrt[3]{5^3} = \sqrt[3]{125} = 5$

3. The product rule: $\sqrt[n]{a} \cdot \sqrt[n]{b} = \sqrt[n]{ab}$

   Example: $\sqrt[3]{a} \cdot \sqrt[3]{b} = \sqrt[3]{ab}$

4. The quotient rule: $\sqrt[n]{\frac{a}{b}} = \frac{\sqrt[n]{a}}{\sqrt[n]{b}}$

   Example: $\sqrt[4]{\frac{a}{b}} = \frac{\sqrt[4]{a}}{\sqrt[4]{b}}$

5. The product rule with coefficients: $a\sqrt{b} \cdot c\sqrt{d} = ac\sqrt{bd}$

   Example: $2\sqrt{7} \cdot 3\sqrt{7} = 6\sqrt{49} = 6(7) = 42$

6. The sum of square roots (same index and radicand): $\sqrt{a}+\sqrt{a}=2\sqrt{a}$

   Example: $\sqrt{8}+\sqrt{8}=2\sqrt{8}=2(2)\sqrt{2}=4\sqrt{2}$

7. Nested radicals: $\sqrt[n]{\sqrt[m]{a}}=\sqrt[nm]{a}$

   Example: $\sqrt[2]{\sqrt[3]{a}}=\sqrt[6]{a}$

8. If $x^2=p$ then $x=\pm\sqrt{p}$
9. Negative signs and exponents:

   $(-a)^2=(-a)\cdot(-a)=a^2$ Example: $(-5)^2=(-5)\cdot(-5)=25$

   $-a^2=-(a\cdot a)=-a^2$ Example: $-5^2=-(5\cdot 5)=-25$

Arithmetic Operations on Square Roots

The **sum of square roots** is not the same as the square root of the sum. You can add or subtract square roots only if the radicand (what's inside the radical sign) is the same. Addition and subtraction have no effect on the radicand. For example:

$\sqrt{2}+\sqrt{2}\neq\sqrt{4}$ $\sqrt{2}+\sqrt{2}=2\sqrt{2}$

$\sqrt{5}-\sqrt{2}\neq\sqrt{3}$ $5\sqrt{2}-3\sqrt{2}=2\sqrt{2}$

Sometimes simplifying the radicals will produce like radicands that can be added or subtracted. Sums of radicands can also be simplified by using the distributive law and collecting like terms. For example,

$3\sqrt{5}+4\sqrt{5}=(3+4)\sqrt{5}=7\sqrt{5}$

Unlike addition, when you multiply square roots the **product of a square root** is equal to the square root of the product. For example:

$\sqrt{5}\times\sqrt{7}=\sqrt{35}$

Multiplication of radicals is used extensively in algebra and calculus so a thorough understanding is important. Below are four examples that will be helpful in solving problems involving radicals.

Example 1: Find the product of $3\sqrt{x}\cdot 2\sqrt{5x}$.

$3\sqrt{x}\cdot 2\sqrt{5x}$

$=(3\cdot 2)\sqrt{x}\sqrt{5x}$ Commutative property

$= 6\sqrt{x \cdot 5x}$ Product rule for radicals

$= 6\sqrt{5x^2}$

$= 6x\sqrt{5}$ Square root of x^2 is x

Example 2: Find the product of $a^2 \cdot \sqrt{a} \cdot a\sqrt{a} \cdot 2a\sqrt{a}$.

$a^2 \cdot \sqrt{a} \cdot a\sqrt{a} \cdot 2a\sqrt{a}$

$= (a^2 \cdot a \cdot 2a)\sqrt{a}\sqrt{a}\sqrt{a}$ Commutative property (grouping)

$= 2a^4\sqrt{a^3}$ Product rule for exponents and radicals

$= 2a^4 \cdot a\sqrt{a}$ Simplify by taking perfect square of a

$= 2a^5\sqrt{a}$ Product rule for exponents

Example 3: Find the product of $\sqrt{3}\left(\sqrt{2} + \sqrt{6}\right)$.

$\sqrt{3}\sqrt{2} + \sqrt{3}\sqrt{6}$ Distributive property

$= \sqrt{6} + \sqrt{18}$ Product rule for radicals

$= \sqrt{6} + \sqrt{2 \cdot 9}$ Find the even root which in this case is 9

$= \sqrt{6} + 3\sqrt{2}$ The square root of 9 is 3 so bring it out in front

Example 4: Find the product of $\left(\sqrt{2} - 3\right)\left(\sqrt{2} + 5\right)$.

$\left(\sqrt{2} - 3\right)\left(\sqrt{2} + 5\right)$

$= \sqrt{2}\sqrt{2} + 5\sqrt{2} - 3\sqrt{2} - 15$ Multiply using FOIL method for binomials (**F**irst **O**uter **I**nner **L**ast terms)

$= \sqrt{4} + 2\sqrt{2} - 15$ Product and subtraction rules for radicals

$= 2 + 2\sqrt{2} - 15$ Simplify

$= -13 + 2\sqrt{2}$

The **quotient of two square roots** is the square root of the quotients of the radicands. For example:

$$\sqrt{\frac{9}{16}} = \frac{\sqrt{9}}{\sqrt{16}} = \frac{3}{4}$$

Simplification of Square Roots

In general, a radical is said to be in simplest form when:

1. The index is as small as possible
2. The radicand has no fractions
3. The denominator of the expression has no radical (rationalize the denominator)
4. Every factor of the radicand has an exponent less than the index

A square root radical expression is simplified when its radicand has no factors that are perfect squares. A **perfect square** is a number that has exact square roots. For example, 16 (4^2), and x^2 are perfect squares. A trinomial such as $a^2+2ab+b^2$ is a perfect square trinomial because its two binomial factors are the same, $(a+b)$.

When simplifying a square root radical, if you do not recognize perfect square factors, factor the radicand into its prime factors. For example:

$$\sqrt{50}=\sqrt{2\cdot 5\cdot 5}=5\sqrt{2}$$

In this example 5×5 is 25 and the square root of 25 is 5, which you bring out in front of the radical.

Simplification of Cube Roots

The simplification of cube roots is similar to the simplification of square roots.

Example 5: Simplify the radical $\sqrt[3]{108x^3y}$

$$\sqrt[3]{108x^3y}=3x\sqrt[3]{4y}$$

1. Determine whether or not the number(s) or variables under the radical consist of a perfect number or in this case a perfect cube. A number is a perfect power when its root can be extracted without leaving a remainder.
2. Pull all perfect cubes out from under the radical. Since $4\cdot 27=108$ you can take the cube root of 27, which is 3. The cube root of 4 is a decimal fraction (1.587) so it is not a perfect cube. The cube root of x^3 is x. You can't take the cube root of y because it is an unknown variable.

Simplification can also be performed using fractional exponents. Using the exponent rule $a^{\frac{m}{n}}=\sqrt[n]{a^m}$, which is division, you can divide m by n and write the quotient as the exponent on the outside of the radical, and the remainder as the exponent on the inside of the radical. For example:

$$\sqrt[3]{y^{26}} = y^{\frac{26}{3}} = y^8\sqrt[3]{y^2}$$

Example 6: Simplify the nested radical $\sqrt[3]{\sqrt{64}}$.

$$\sqrt[3]{\sqrt{64}} = \left[(64)^{\frac{1}{2}}\right]^{\frac{1}{3}} = 8^{\frac{1}{3}} = \sqrt[3]{8} = 2$$

An alternative method is shown below:

$\sqrt[3]{\sqrt{64}} = \sqrt[6]{64} = 2$ Nested radical rule: $\sqrt[n]{\sqrt[m]{a}} = \sqrt[nm]{a}$

Check: $2^6 = 64$

Example 7: Simplify the radical $\sqrt[8]{16b^{12}}$.

$\sqrt[8]{16b^{12}} = \sqrt[8]{2^4 \cdot b^8 \cdot b^4}$ Convert the radicand to even powers

$= 2^{\frac{4}{8}} \cdot b^{\frac{8}{8}} \cdot b^{\frac{4}{8}}$ Convert to fractional exponent form

$= 2^{\frac{1}{2}} \cdot b \cdot b^{\frac{1}{2}}$ Simplify fractional exponents

$= b\sqrt{2b}$ Write in radical form. Remember that $b^{\frac{1}{2}}$ is the square root of b^1.

Example 8: Reduce $\sqrt[4]{16b^4y^3}$ to its simplest form.

$$\sqrt[4]{16b^4y^3} = \sqrt[4]{2^4b^4y^3} = 2b\sqrt[4]{y^3}$$

The index 4 is as small as possible, the radicand has no fraction, there is no radical in the denominator of the expression, and the radicand y^3 has no factor which is a 4th power of y.

Additional radical simplification and manipulation techniques include the following:

A factor of the radicand can be removed from under the radical sign when the factor is an exact power of the index.

Example 9: Simplify the radical $\sqrt[3]{81}$.

$\sqrt[3]{81} = \sqrt[3]{3 \cdot 27} = 3\sqrt[3]{3}$ The cube root of 27 is 3 so you bring it in front of the radical

A factor in the coefficient of a radical can be moved under the radical sign by raising the factor to the power of the index.

Example 10: Move the coefficient of the radical under the radical for $2\sqrt[3]{4}$.

$$2\sqrt[3]{4} = \sqrt[3]{(2^3)4} = \sqrt[3]{8 \cdot 4} = \sqrt[3]{32}$$

A radical with a fractional radicand can be reduced to a fraction whose denominator has no radical by multiplying the numerator and denominator by one. This is one method for rationalizing the denominator, which is examined in further detail below.

To simplify the square root of a fraction extract the root of both the numerator and denominator separately. This is an application of the quotient rule for radicals $\sqrt[n]{\frac{a}{b}} = \frac{\sqrt[n]{a}}{\sqrt[n]{b}}$. For example, $\sqrt{\frac{49}{64}} = \frac{\sqrt{49}}{\sqrt{64}} = \frac{7}{8}$.

The square root of a decimal has half as many decimal places as does the decimal itself. For example, $\sqrt{.0081} = .09$, $\sqrt{1.5625} = 1.25$.

Rationalizing the Denominator

When you **rationalize the denominator** you remove the radical term from the denominator. This is accomplished by multiplying a radical fraction by the rationalizing factor. The rationalizing factor is the factor by which the numerator and the denominator of a radical fraction are multiplied to make the denominator rational. The product of any number multiplied by itself will be a perfect square. There are two methods for rationalizing the denominator as shown below.

Example 11: Rationalize the radical $\frac{\sqrt{8}}{\sqrt{3}}$.

$$\frac{\sqrt{8}}{\sqrt{3}} = \frac{\sqrt{8}}{\sqrt{3}} \cdot \frac{\sqrt{3}}{\sqrt{3}} = \frac{\sqrt{24}}{3} = \frac{\sqrt{4 \cdot 6}}{3} = \frac{2\sqrt{6}}{3}$$

In this example you multiply the numerator and denominator by the radical in the denominator. When you multiply a radical times a radical, it has the same effect as removing the radical sign.

Example 12: Reduce the fractional radicand so that there is no radical in the denominator for $\sqrt{\frac{3}{7}}$.

$$\sqrt{\frac{3}{7}}=\sqrt{\frac{3\cdot 7}{7\cdot 7}}=\sqrt{\frac{21}{49}}=\frac{\sqrt{21}}{\sqrt{49}}=\frac{\sqrt{21}}{7}$$

Multiplying by $\frac{7}{7}$ is the same as multiplying by 1

Finding Square Roots

One way to find approximate square roots is with the divide and average method for finding square roots.

Algorithm: To find a square root with the divide and average method:

1. Approximate the square root of the number
2. Divide the number x by the approximate square root
3. Take the average of the approximate square root and the quotient

The formula for this method is as follows:

$$x_1=\frac{1}{2}\left(x_0+\frac{2}{x_0}\right)$$

Where x_0 is the first approximation and x_1 is the first result

Example13: What is the approximate square root of 2?

First guess 1.5 or $1.5^2=2.25$.

$$x_1=\frac{1}{2}\left(x_0+\frac{2}{x_0}\right)$$

$$x_1=\frac{1}{2}\left(1.5+\frac{2}{1.5}\right)$$

$$=\frac{1}{2}(1.5+1.33)$$

$$=\frac{1}{2}(2.83)$$

$$= 1.41 \rightarrow 1.41^2 = 2.002 \quad \text{1.41 becomes your second guess}$$

$$x_2 = \frac{1}{2}\left(x_1 + \frac{2}{x_1}\right)$$

$$x_2 = \frac{1}{2}\left(1.41 + \frac{2}{1.41}\right)$$

$$= \frac{1}{2}(1.41 + 1.418)$$

$$= \frac{1}{2}(2.828)$$

$$= 1.414 \rightarrow 1.414^2 = 1.999$$

The actual square root of 2 is irrational (non-terminating decimal) and is equal to 1.414213562 to nine decimal places. In this case, just two trials produces a very good approximation.

Summary

An exponent is a number placed at the right of and above a number or a symbol. The value assigned to the symbol with this exponent is called a power and indicates the power taken or how many times the number or symbol is multiplied by itself. A fractional exponent is an exponent expressed as fraction that indicates a root of an expression. The numerator indicates the power to which the base is to be raised, and the denominator, the root which is to be extracted of that power. A negative fractional exponent indicates the reciprocal of the expression.

A logarithm is nothing more than an exponent. A radical is an expression used to indicate the root of a number. The root of a number is one of the equal factors of a number. A radical is said to be in simplest form when: the index is as small as possible, the radicand has no fractions, the denominator of the expression has no radical (rationalize the denominator), and every factor of the radicand has an exponent less than the index. Finally, the divide and average method for finding square roots is reviewed.

When you rationalize the denominator you remove the radical term from the denominator. This is accomplished by multiplying a radical fraction by the rationalizing factor. The rationalizing factor is the factor by which the numerator and the denominator of a radical fraction are multiplied to make the denominator rational.

Number Systems

When numbers were examined earlier, the concept of **place value** was defined as a numeration system in which a real number is represented by an ordered set of numerals in arithmetic or characters such as letters in algebra, where the value of a numeral or character depends on its position. Each position is related to the next by a constant multiplier called the base of that numeral system.

A group of symbols that can be used to express numbers is called a **numeration system**. The size of the groupings in a numeration system is the **number base**, also referred to as base or radix. The symbols of a numeration system are called **numerals**. Each digit in a positional numeral conveys two things: its **face value**, which is the inherent value of the particular symbol used, and its place value, which is the power of the base associated with the position that the digit occupies.

To work with number systems it is helpful to use expanded notation. Numbers can be written in expanded notation which shows the place value of each digit. In the example below, 345 is written in expanded notation with the last example using exponents.

$$300+40+5$$
$$(3\times100)+(4\times10)+(5\times1)$$
$$(3\times10^2)+(4\times10^1)+(5\times10^0)$$

Since computers operate using binary operations and hexadecimal values it is important to understand how these number systems work and how to convert numbers from decimal to binary and from binary to decimal, and from decimal to hexadecimal and from hexadecimal to decimal. In an exponential expression, the **base** is the number that is used as factor a given number of times. Up to now we have worked only with numbers of the base 10 or decimal numbers. The **binary number system** is a number system whose base is 2. The binary number system only uses the symbols 0 and 1 whereas the decimal number system used the symbols 0, 1, 2, 3, 4, 5, 6, 7, 8, 9. The place values of the digits in the binary number system are powers of 2. For example, $2^0, 2^1, 2^2, 2^3$.

Example 1: Convert 11100110 to decimal form.

$$
\begin{aligned}
11100110 &= (1\times 2^7)+(1\times 2^6)+(1\times 2^5)+(0\times 2^4)+(0\times 2^3)+(1\times 2^2)+(1\times 2^1)+(0\times 2^0) \\
&= 128+64+32+0+0+4+2+0 \\
&= 230
\end{aligned}
$$

Example 2: Convert 230 to binary form.

$$
\begin{array}{l l}
2\,\underline{|230} & \\
2\,\underline{|115} & 0 \\
2\,\underline{|57} & 1 \\
2\,\underline{|28} & 1 \\
2\,\underline{|14} & 0 \\
2\,\underline{|7} & 0 \\
2\,\underline{|3} & 1 \\
2\,\underline{|1} & 1 \\
\ \ \underline{|0} & 1
\end{array}
$$

Divide 230 by 2 and write the remainders at the side.

In this example, you start by dividing 230 by 2 which equals 115 with a remainder of 0. The remainders are shifted down one place. Next, you divide 115 by 2 which equals 57 with a remainder of 1. Continue until the last dividend is zero. To write in binary form, write the remainders from the bottom to the top: 230 = 11100110.

The **hexadecimal number system** is a number system whose base is 16. The hexadecimal system uses the numbers 0 – 9 and the letters A – F. In the hexadecimal system additional digits are needed to go beyond the 10 digits of base ten. The letters A, B, C, D, E, and F represent the hexadecimal digits for the numbers ten through fifteen. Where A = 10, B = 11, C = 12, D = 13, E = 14, and F = 15.

Example 3: Convert 20,387 from base 10 to hexadecimal.

16 | 20387		
16 | 1274	3	3
16 | 79	10	A
16 | 4	15	F
| 0	4	4

Divide 20,387 by 16 and write the remainders at the side. The process is the same as converting decimal to binary except you divide by sixteen instead of two. The remainders are written in their hexadecimal equivalent next to the decimal remainders. Write the remainders from the bottom to the top: 20,387 = 4FA3.

Example 4: Convert FA5 base 16 to base ten.

$$\begin{aligned} FA5 &= (15\times16^2)+(10\times16^1)+(5\times16^0) \\ &= 3840+160+5 \\ &= 4005 \end{aligned}$$

Multiply each hex value by the respective place value and add the products. The letter F represents the hexadecimal digit 15 and the letter A represents the hexadecimal digit 10.

Summary

Place value is a numeration system in which a real number is represented by an ordered set of characters where the value of a character depends on its position. A group of symbols that can be used to express numbers is called a numeration system. The size of the groupings in a numeration system is the number base, or base. The symbols of a numeration system are called numerals. Each digit in a positional numeral conveys two things: its face value, which is the inherent value of the particular symbol used, and its place value, which is the power of the base associated with the position that the digit occupies.

The binary number system is a number system whose base is 2. The binary number system only uses the symbols 0 and 1. The hexadecimal number system is a number system whose base is 16. The letters A, B, C, D, E, and F represent the hexadecimal digits.

The Number Line and Coordinate Systems

Before the introduction to algebra there are several topics that should be understood before proceeding. These topics include the number line and coordinate systems, the order of operations, the usage of subscripts and superscripts in mathematics, and scientific notation.

Coordinate Systems

Definition: A **coordinate system** is any method by which a number or set of numbers is used to represent a point, line, or geometric object. The set of numbers are called coordinates of the point. A coordinate system may be one, two or three-dimensional. A one dimensional coordinate system has one axis (symmetrical line). An example of a one dimensional coordinate system is the number line. A two dimensional coordinate system has two axes (typically labeled x, and y) and is used to locate a point in a plane. As you may remember from geometry, a **point** represents position and has no length, width, or height. A **plane** is a flat (no height) two dimensional surface, which contains at least three non-collinear (not on the same line) points and extends to infinity in all directions. The Cartesian or rectangular coordinate system and the polar coordinate system are examples of two dimensional coordinate systems. A three dimensional coordinate system has three axes (x, y, z) and is used to locate a point in space.

Mathematical Methods - Basic Mathematics will examine the number line and the two dimensional coordinate system. More advanced coordinate systems such as three dimensional coordinate systems and polar coordinate systems will be examined in *Mathematical Methods - Geometry and Trigonometry*.

The Number Line

Definition: The **number line** is a line, each point of which is associated with the number that gives its distance from a fixed point (the origin or zero) on the line. A number line is sometimes called a coordinate line. The number associated with a given point on the line is called the coordinate of that point. To find the distance between any two points on a number line subtract one coordinate from the other and disregard the sign of the distance. Distance is an unsigned number or absolute value. To indicate directed distance (positive or negative direction) use the sign.

Figure 1: The number line

The number line can be used to display addition and subtraction. In the example below, the number line is used to find the sum of +3 and +4. First count three units to the right of zero to point A. From this point, count four more units to the right to point B which is seven units to the right of zero. Therefore, 7 is the sum.

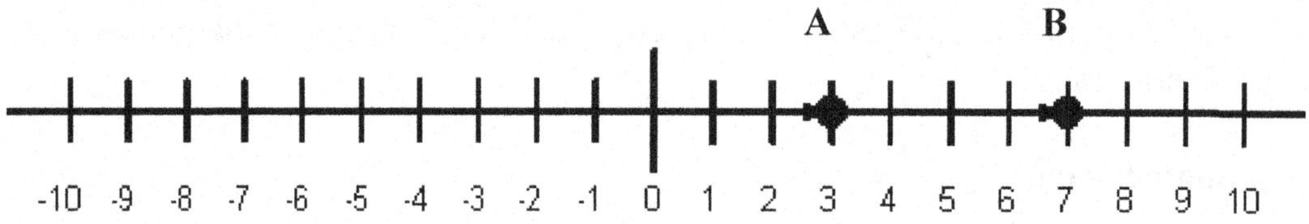

To subtract -3 from -7, count seven units to the left to point A. Since $-(-3)$ is $+3$ from this point move three units to the right to point B which is -4 units to the left of zero. Therefore, -4 is the difference or remainder.

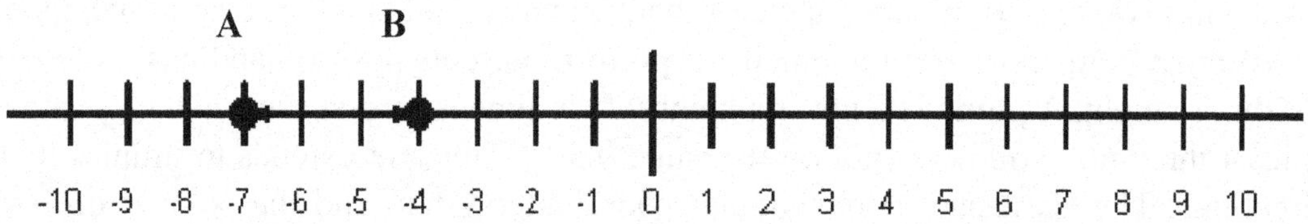

The Cartesian or Rectangular Coordinate System

The **rectangular coordinate system** also referred to the **Cartesian coordinate system** after the French mathematician and philosopher Rene Descartes is a coordinate system used to locate a point with reference to two perpendicular lines. **Perpendicular** lines are lines that form 90 degree right angles. A point in the plane is located in terms of its distance from each of the two perpendicular lines, called axes (plural for axis). The rectangular coordinate system is made up of two lines, one horizontal line (x-axis) and one vertical line (y-axis). The two lines are perpendicular to each other and intersect at a midpoint called the origin. The horizontal line or the **abscissa** is called the *x*-axis and represents the distance of the point from the origin on the *x*-axis. Like the number line, the *x*-axis is positive to the right of the origin and negative to the left of the origin. The vertical line or the **ordinate** is called the *y*-axis and represents the distance of the point from the origin on the *y*-axis. The *y*-axis is positive above the *x*-axis and negative below the *x*-axis.

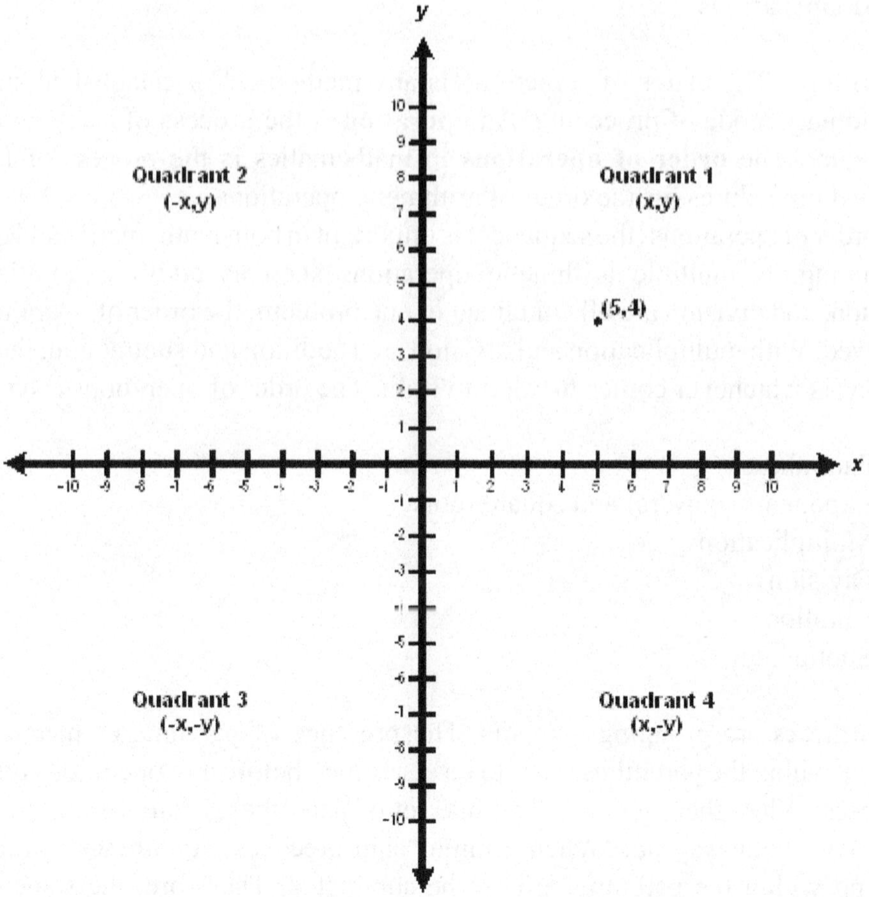

Figure 2. Cartesian coordinate system

Figure 2 above shows the structure of the Cartesian coordinate system. As stated above, the system is made up of two perpendicular lines called the *x* and *y*-axis. The figure shows each axis labeled from -10 to +10. The axes go to infinity in both the positive and negative direction as shown by the arrows at the end of the axes. The intersecting axes separate the plane into four parts called **quadrants**. They are labeled quadrants 1, 2, 3, and 4 in counterclockwise order. The **origin** is the point where the two axes intersect and is represented by the coordinates (0, 0). The coordinates (*x*, *y*) of all points in the Cartesian plane are ordered pairs. An **ordered pair** is two elements in a definite sequence. For example, the number elements 5 and 4 form two different ordered pairs (5, 4) and (4, 5). The elements are the same but the order is opposite and each represents a different point. The (*x*, *y*) coordinate signs are shown for each quadrant. In quadrant 1, both x and y are positive (x, y), in quadrant 2, x is negative and y is positive $(-x, y)$, in quadrant 3, both x and y are negative $(-x, -y)$ and in quadrant 4, x is positive and y is negative $(x, -y)$. The point (5, 4) is shown in the first quadrant. This coordinate has an x value that is 5 units to the right of the origin in the positive direction and a y value that is 4 units in the positive vertical direction from the origin.

Order of Operations

Definition: The **order** of something is any methodical or established succession or a customary mode of procedure. An **operation** is the process of carrying out rules of procedure. The **order of operations** in mathematics is the process of following established procedures for the order of arithmetic operations.

The order of operations (the sequence) is important in both mathematics and computer programming. If multiple arithmetic operations such as addition, multiplication, subtraction, and division are all contained in one problem, the order of operations must be followed. With multiplication and division, and addition and subtraction, the order of operations is whichever comes first left to right. The order of operations is as follows:

1. Parentheses
2. Exponents (powers) and square roots
3. Multiplication
4. Division
5. Addition
6. Subtraction

Parentheses are grouping symbols. The presence of parentheses means that the operations within the parentheses are to be performed before any operations outside the parentheses. When there is more than one set of parentheses start with the innermost set and work your way out. When a minus sign precedes parentheses, it means that everything within the parentheses is to be subtracted. Therefore, the same rule that applies to the subtraction of signed numbers applies. Change every sign within the parentheses to its opposite sign and add. If the expression enclosed in the parentheses is preceded by a plus sign, remove the parentheses with no sign change.

The order of operations can be remembered by using the acronym PEMDAS (**P**lease **E**xcuse **M**y **D**ear **A**unt **S**ally) which stands for **P**arentheses, **E**xponents, **M**ultiplication, **D**ivision, **A**ddition, and **S**ubtraction.

Example 1: Calculate the result for the following expression:

$10-3\times6+10^2+(6+1)\times4$

$=10-3\times6+10^2+(7)\times4$ Parentheses first

$=10-3\times6+100+(7)\times4$ Exponents second

$=10-18+100+28$ Multiplication

$=-8+100+28$ Addition / subtraction left to right

$= 92 + 28$

$= 120$

Subscripts and Superscripts

Subscripts and superscripts are used extensively in higher level mathematics so an understanding of their definitions and usage are provided. A **subscript** is that which is written below something else; a subscript is a small number or letter written below and to the right or left of a letter as a mark of distinction or as part of an operative symbol. Subscripts are often used to distinguish one symbol from another representing the same type of quantity. For example, v_1 and v_2 are used to represent two different velocities in the same problem. The symbols a_1, a_2 denote constants; $(x_0, y_0), (x_1, y_1)$ denote two different coordinates of fixed points. In combinatorial analysis ${}_nC_r$ denotes the number of possible combinations of *n* things *r* at a time. Subscripts generally don't do anything mathematically; they simply allow you to tell the difference between two or more items that look alike.

A **superscript** is that which is written above something else. A superscript is a number or symbol written above and to the right or left of a letter or number, usually denoting an exponent (power) or a derivative (the instantaneous rate of change of a function with respect to a variable). For example, x^2 is *x* to the second power; in calculus x'' uses the prime superscript as a symbol to represent the second derivative of *x*.

Scientific Notation

Scientific notation is a notation for writing very large or very small numbers by using powers of ten. In scientific notation, a number is written in the form $A \times 10^n$ where *A* is a number greater than or equal to 1 and less than 10, and the exponent *n* is a positive or negative integer.

Example 1: Write the number 4853 in scientific notation.

A has to be greater than or equal to 1 and less than 10. Therefore, 4853 would be written in scientific notation as 4.853×10^3 which is 4.853 multiplied by 3 factors of 10 or one thousand as shown below:

$4.853 \times 10 \times 10 \times 10 = 4853$

Example 2: Write the number 0.00568 in scientific notation.

The number 0.00568 would be written as 5.68×10^{-3} which is 5.68 divided by 3 factors of 10 as shown below:

$$5.68 \times 10^{-3} = \frac{5.68}{10 \times 10 \times 10} = \frac{5.68}{1000} = .00568$$

To add and subtract two numbers written in scientific notation it is necessary to express both numbers to the same power of ten. After adding and subtracting it may be necessary to shift the decimal point to express the result in scientific notation.

Example 3: Add the following and express the sum in scientific notation:

$$\left(9.42 \times 10^{-2}\right) + \left(7.6 \times 10^{-3}\right)$$

The decimal point can be shifted in either number in order to convert both to the same power of 10. To get both numbers to 10^{-2}, shift the decimal point one place to the left and add 1 to the exponent in the expression 7.6×10^{-3}.

$$7.6 \times 10^{-3} = 0.76 \times 10^{-2}$$

Now the two numbers can be added.

$$\begin{aligned} \left(9.42 \times 10^{-2}\right) + \left(7.6 \times 10^{-3}\right) &= (9.42 + 0.76) \times 10^{-2} \\ &= 10.18 \times 10^{-2} \\ &= 1.018 \times 10^{-1} \end{aligned}$$

Since 10.18 is not between 1 and 10, shift the decimal point to the left and reduce the exponent by one to express the final result in scientific notation.

To multiply two numbers in scientific notation, first multiply the two powers of 10 by adding their exponents (see rules of exponents). Then multiply the remaining factors. To divide two numbers in scientific notation move any power of 10 in the denominator to the numerator and reverse the sign of the exponent. Reversing the sign of the power in the denominator is the negative exponent rule. The division rule for exponents could also be used and the result would be the same. After multiplying the two powers of 10, carry out the indicated division.

Example 4: Multiply the following and express the product in scientific notation:

$$\left(6.3 \times 10^{2}\right) \times \left(2.4 \times 10^{5}\right)$$

$$\left(6.3 \times 10^{2}\right) \times \left(2.4 \times 10^{5}\right) = (6.3 \times 2.4) \times 10^{7}$$ Multiplication rule: $a^n a^m = a^{n+m}$

$$= 15.12 \times 10^7$$ Multiply factors

$$= 1.512 \times 10^8$$ Shift the decimal point to the left and increase the exponent by 1.

Example 5: Divide the following and express the quotient in scientific notation:

$$\frac{6.4 \times 10^2}{2.0 \times 10^5}$$

$$\frac{6.4 \times 10^2}{2.0 \times 10^5} = \frac{6.4}{2.0} \times 10^2 \times 10^{-5}$$

Move the power of 10 in the denominator to the numerator and change its sign to the opposite.

Negative exponent rule: $a^{-n} = \frac{1}{a^n}, a^{-1} = \frac{1}{a}$

$$= \frac{6.4}{2.0} \times 10^{-3}$$

$$= 3.2 \times 10^{-3}$$ or .00032

You could also use the exponent division rule $\frac{a^n}{a^m} = a^{n-m}$. Where $a^n = 10^2$ and $a^m = 10^5$, and $a^{n-m} = 10^2 - 10^5 = 10^{-3}$.

$$\frac{6.4 \times 10^2}{2.0 \times 10^5} = 3.2 \times 10^{-3}$$

Summary

The number line is a line, each point of which is associated with the number that gives its distance from a fixed point (the origin or zero) on the line. A coordinate system is any method by which a number or set of numbers is used to represent a point, line, or geometric object. The set of numbers are called coordinates of the point. A two dimensional coordinate system has two axes (typically labeled *x*, and *y*) and is used to locate a point in a plane. A three dimensional coordinate system has three axes (*x*, *y*, *z*) and is used to locate a point in space.

The rectangular coordinate system also referred to the Cartesian coordinate system after the French mathematician and philosopher Rene Descartes is a coordinate system used to locate a point with reference to two perpendicular lines. A point in the plane is located in terms of its distance from each of the two perpendicular lines, called axes (plural for axis). The rectangular coordinate system is made up of two lines, one horizontal line (*x*-axis) and one vertical line (*y*-axis).

The order of operations in mathematics is the process of following established procedures for the order of arithmetic operations. The order of operations can be remembered by using the acronym PEMDAS (**P**lease **E**xcuse **M**y **D**ear **A**unt **S**ally) which stands for **P**arentheses, **E**xponents, **M**ultiplication, **D**ivision, **A**ddition, and **S**ubtraction.

Scientific notation is a notation for writing very large or very small numbers by using powers of ten. In scientific notation, a number is written in the form $A \times 10^n$ where A is a number greater than or equal to 1 and less than 10, and the exponent n is a positive or negative integer.

Introduction to Algebra

The following is a brief introduction to algebra, which will provide the basics of equations. Equations are useful in solving problems discussed in the chapters on Ratio, Proportion, Percentage and Variation, and Sequences, Summation and Series. Algebra will be reviewed in detail in Volume 2 of the Mathematical Methods series *Mathematical Methods – Algebra.*

Definition: Algebra is the branch of mathematics in which the operations of arithmetic are generalized by the use of letters to represent quantities. Algebra is the abstraction of arithmetic. The basic task of elementary algebra is the solution of polynomial equations obtained by the operations of addition, subtraction, multiplication, division, raising to a power, and extracting a root. If only the first four operations are performed a finite number of times the resulting numbers, functions, or equations are rational. When fractional exponents and the extraction of roots are involved the quantities can be irrational and have complex solutions (imaginary numbers).

History: The earliest known treatise which could be called algebraic is the Ahmes Papyrus written around 1600 B.C. now in the British Museum. The only Greek to write on algebra was Diophantus, who is thought to have lived in the mid third century A.D. and was probably the first to use a symbol to represent an unknown quantity. Diophantus, often known as the 'father of algebra', is best known for his *Arithmetica*, a work on the solution of algebraic equations and the theory of numbers.

Algebra was further developed by Al-Khwarizmi in his treatise called *al-Kitab al-mukhtasar fi hisab al-jabr w'al-muqabala* or *The Compendious Book on Calculation by Completion [or Restoring] and Balancing*. This book has an explanation of the solutions to quadratic and linear equations of six varieties. *Al-jabr* refers to the process of moving a subtracted quantity to the other side of an equation; *al-muqabala* involves subtracting equal quantities from both sides of an equation. *Al-Kitab al-mukhtasar fi hisab al-jabr w'al-muqabala* was translated into Latin in 1140 A.D. as *Liber algebrae et almucabala*, from which we have the word "algebra" for the whole process of solving equations.

Applications: Algebra is the foundation of all higher level mathematics. The main branches of algebra are elementary algebra, linear algebra, abstract algebra, and Boolean algebra. In elementary algebra, one learns to calculate with variables instead of just the numbers of arithmetic and to solve polynomial equations. Abstract algebra is more advanced where the variables x and y are used to represent elements other than just numbers, such as polynomials and matrices. Linear algebra is the branch of mathematics that involves the study of systems of linear equations, vector spaces, linear mappings, and linear, bilinear, and quadratic functions on vector spaces. Boolean algebra is the algebra of subsets of a collection of subsets of some set. Whereas elementary algebra is based on the numeric operations multiplication xy, addition $x + y$, and negation $-x$, Boolean algebra is based on logical counterparts to those operations, namely conjunction (AND), disjunction (OR), and complement or negation (NOT).

Algebra – Solving Equations

There are two types of equations or statements in mathematics, the identity and the conditional equation. An **identity** is a statement of equality, usually denoted by $\equiv$, which is true for *all* values of the variables, with the exception of values of the variables for which each member of the statement of equality does not have meaning for example, division by zero for example. The = sign is commonly used in place of the identity sign $\equiv$. For example, $(x+y)^2 = x^2 + 2xy + y^2$ is an identity.

A **conditional equation** is true only for *some* values of the variables involved. For example, $x + 2 = 5$ is a true statement only when $x = 3$. A **solution** (or root) of a conditional equation is a value of the variable (or set of variables if there is more than one) for which the equation is a true statement.

There are five concepts that are important in the understanding of equations. Equations are made up of terms, like terms, variables, coefficients, and expressions.

A **term** is a unitary (single) or compound expression connected with another by a plus or minus sign; an element of a fraction or proportion, or of a series or sequence; or any distinct quantity contained in a polynomial. Stated more simply, a **term** is the part of an algebraic expression between a plus and a minus sign, which only contains algebraic symbols and numbers. For example, $4x^2 + 2x$ are two terms. **Like terms** are terms that contain the same variables with each variable of the same kind being raised to the same power. For example, $4x$ and $16x$ are like terms, and $2x^2yz$ and $5x^2yz$ are like terms.

A **variable** is something that is changeable or has the capacity of varying or changing. Letters from the end of the alphabet, such as *v*, *w*, *x*, *y*, *z* are generally used for a quantity that is variable. In algebra, a variable generally represents an unknown quantity that you are trying to find.

A **constant** is something that is fixed or remains unchanged or invariable. Letters from the beginning of the alphabet, such as *a*, *b*, *c*, *d*, *e* are generally used for a quantity that is constant in any one problem.

A **coefficient** is a number or letter placed before or in front of a letter or quantity, known or unknown, to show how many times the latter is to be taken. For example, $6x$; bx; here 6 and b are coefficients of x. For $5aby$, $5ab$ is the coefficient because y is a variable, and a and b are constants.

An **expression** is a quantity made up of letters, numerals, and algebraic symbols. The parts of an expression that are connected by plus and minus signs are the terms of an expression. The sign immediately preceding the term is the sign of that term. If there is no sign preceding the term it is understood to be positive. The expression $3x-2y$ has two terms, $3x$ and $-2y$. If an algebraic expression has one term, it is called a monomial (mono means one). If the expression has two terms it is called a binomial (bi means two), and if the expression has three terms it is called a trinomial (tri means three).

Example: What are the terms, variables, constants, and coefficients of the following equation?

$$x^2-7x+6$$

There are three terms x^2, $-7x$, and 6. There is one variable x. There is one constant, 6. The coefficient of x^2 is understood to be 1. The coefficient of $-7x$ is -7, therefore there are two coefficients.

A **solution** of an equation is a number replacement for the variable that makes the equation a true statement. A solution is sometimes referred to as a root. To solve an equation means to isolate the variable (so the variable is by itself), which in turn means to get the variable you are solving for to one side of the equation by itself. The first rule is that anything you do to one side of the equation you have to do to the other side of the equation. The equation must stay in balance. If you perform the same operations to both sides of the equal sign, the equation will stay balanced. To undo an operation apply the inverse operation to both sides. When you raise both sides of an equation to an even power you need to check your work or you could end up with extraneous solutions.

An **extraneous solution** is a number obtained in the process of solving an equation, which is not a root of the equation. It is generally introduced by squaring or clearing denominators in the original equation. For example, the equation $\frac{x^2-3x+2}{x-2}=0$ has only one root, 1; but if you multiply through by $x-2$ to clear the denominator the resulting equation has a root of 2. You have to check your work by substituting the value for the variable to verify that the equation is true.

You can take the even root of both sides only when you know that both sides are positive. When you know they are positive and take an even root, you need to add the plus or minus sign to the side of the even root.

On multistep equations there are problems where it is better to move whole terms to one side or the other to get all the terms with the variable on one side. Once you get every term on one side you can combine terms or factor to further isolate the variable. Before you transpose factored terms you need to remove the parentheses.

For many equations you can use the certified method **CeRTiFieD** – **C**lear all denominators (multiply the entire equation by the common denominator of all fractions), **R**emove all parentheses, **T**ranspose the terms to get all like terms on one side (collect like terms), **F**actor and **D**ivide by the terms that will leave the desired variable. In some cases when you have like terms in the numerator but not the denominator it is cleaner to divide both sides of the expression to simplify it. If there are fractions that can be reduced, reduce before beginning the certified method.

For an equation such as $y = 2x + 8$ the equation expresses y in terms of x. It is an equation that has y alone on one side and has an algebraic expression that includes x on the other side. When you solve an equation like this for x, you end up with an equation that expresses x in terms of y, where in this case $x = \frac{y-8}{2}$. When something is 'in terms of something' it means with respect to, or in relation to. Respect is defined as a relation or reference to a particular thing or situation and 'with respect to' means with reference to or in relation to.

Summary

Algebra is the branch of mathematics in which the operations of arithmetic are generalized by the use of letters to represent quantities. Algebra is the abstraction of arithmetic.

There are two types of equations or statements in mathematics, the identity and the conditional equation. An identity is a statement of equality, usually denoted by $\equiv$, which is true for *all* values of the variables, with the exception of values of the variables for which each member of the statement of equality does not have meaning for example, division by zero.

A conditional equation is true only for *some* values of the variables involved. For example, $x+2=5$ is a true statement only when $x=3$. A solution (or root) of a conditional equation is a value of the variable (or set of variables if there is more than one) for which the equation is a true statement.

Equations are made up of terms (a single or compound expression connected with another by a plus or minus sign), like terms, variables (something that is changeable or has the capacity of varying or changing), coefficients (a number or letter placed before or in front of a letter or quantity), and expressions (a quantity made up of letters, numerals, and algebraic symbols).

A solution of an equation is a number replacement for the variable that makes the equation a true statement. A solution is sometimes referred to as a root. To solve an equation means to isolate the variable (so the variable is by itself), which in turn means to get the variable you are solving for to one side of the equation by itself. When solving equations you need to check your answers to make sure you do not get extraneous solutions. An extraneous solution is a number obtained in the process of solving an equation, which is not a root of the equation. It is generally introduced by squaring or clearing denominators in the original equation.

When solving equations, anything you do to one side of the equation you must do to the other side of the equation. The equation must stay in balance. For many equations you can use the certified method **CeRTiFieD** – **C**lear all denominators (multiply the entire equation by the common denominator of all fractions), **R**emove all parentheses, **T**ranspose the terms to get all like terms on one side (collect like terms), **F**actor and **D**ivide by the terms that will leave the desired variable.

Algebra – Linear Equations

Linear or First Degree Equations

Definition: A **linear equation** is an equation that may be expressed in the form $ax + by = c$, where *a*, *b*, and *c* are any real numbers and $a \neq 0$. No term is higher than the **first degree,** which means a variable in a linear equation has an exponent of 1; that is $x = x^1$. First degree equations are called linear because its graph is always a straight line. A linear equation in one variable usually has only one solution. The main forms of linear equations are shown and will be examined in greater detail below.

Linear Equations

1. General form of a linear equation: $Ax + By + C = 0$
2. Slope formula: $m = \dfrac{\textit{change in } y}{\textit{change in } x} = \dfrac{y_2 - y_1}{x_2 - x_1}$
3. Slope intercept form: $y = mx + b$, where *m* is the slope and *b* is the *y*-intercept
4. Point-slope form (used to find the equation of a line): $y - y_1 = m(x - x_1)$
5. Horizontal line parallel to the *x*-axis: $y = b$
6. Vertical line parallel to the *y*-axis: $x = a$

Applications: Linear equations are used for modeling and calculations involving rates, for linear regression analysis in statistics, which is a technique for estimating the straight line relationship between two variables, and in calculus for finding the equation of a line based on the derivative and a point on the line. Slope can be considered as the average rate of change. To find the average rate of change between two data points you find the slope of the line that passes through the two points.

Algorithm: To solve all first degree equations:

1. Simplify by first removing parentheses if they exist. Further simplify by clearing all fractions by multiplying the entire equation by the least common denominator of all fractions.
2. Simplify both sides of the equation by collecting or grouping like terms and by adding or subtracting like terms
3. Isolate the variable by transposing or moving all terms with variables to one side, and everything else to the other side.
4. Divide the equation by the coefficient of the variable to isolate the variable and make the coefficient of the variable equal 1.
5. Check the solution by replacing the variable with the solution in the original equation.

Example 1: Solve the linear equation $3x+4=13$.

$3x+4=13$	
$3x+4-4=13-4$	Subtract 4 from both sides of the equation
$3x=9$	Result of subtracting 4 from both sides
$\frac{3x}{3}=\frac{9}{3}$	Divide both sides by 3 to isolate the variable x
$x=3$	The solution is $x=3$

Check:

$3(3)+4=13$	
$13=13$	Both sides of the equation are 13 so $x=3$ is correct

Example 2: Solve the following linear equation for x, $\frac{2x}{3}-\frac{1}{2}=5-\frac{x}{4}$.

$\frac{2x}{3}-\frac{1}{2}=5-\frac{x}{4}$	
$12\left(\frac{2x}{3}\right)-12\left(\frac{1}{2}\right)=12\left(5-\frac{x}{4}\right)$	Clear the fractions by multiplying both sides of the equation by the LCD 12
$8x-6=60-3x$	

$8x+3x-6=60-3x+3x$	Add $3x$ to both sides of the equation to isolate the variable
$11x-6=60$	
$11x-6+6=60+6$	Add 6 to both sides of the equation
$11x=66$	Divide both sides of the equation by 11 to make the coefficient of x equal to 1

$$\frac{11x}{11}=\frac{66}{11}$$

$$x=6$$

Check:

$$\frac{2x}{3}-\frac{1}{2}=5-\frac{x}{4}$$

$$\frac{2(6)}{3}-\frac{1}{2}=5-\frac{6}{4}$$

$$\frac{12}{3}-\frac{1}{2}=\frac{20}{4}-\frac{6}{4}$$

$$\frac{24}{6}-\frac{3}{6}=3\frac{1}{2}$$

$$3\frac{1}{2}=3\frac{1}{2}$$

$3\frac{1}{2}=3\frac{1}{2}$ is true so the solution is $x=6$

Example 3: Solve the following linear equation for x,

$2x-2=-3x+3$	
$2x-2+2=-3x+3+2$	Add 2 to both sides of the equation
$2x=-3x+5$	Result of adding 2 to both sides of the equation
$2x+3x=-3x+3x+5$	Add $3x$ to both sides of the equation
$5x=5$	Result of adding $3x$ to both sides of the equation
$\frac{5x}{5}=\frac{5}{5}$	Divide both sides of the equation by 5 to isolate the variable x
$x=1$	

Check:

$$2x-2=-3x+3$$
$$2(1)-2=-3(1)+3$$
$$0=0$$

$0=0$ is true so the solution is $x=1$

Two additional concepts that are important in algebra and calculus are intercepts and slope. The **intercept** of a straight line, curve, or surface on an axis of coordinates is the distance from the origin to the point where the line, curve, or surface cuts or crosses the x or y axis. The ***x*-intercept** $(x, 0)$ is the point at which the graph intersects the x-axis. It will always have a y-coordinate of zero. A horizontal line that is not the x-axis will not have an x-intercept. The ***y*-intercept** $(0, y)$ is the point at which the graph intersects the y-axis. It will always have an x-coordinate of zero. A vertical line that is not the y-axis will not have a y-intercept.

Generally, a **slope** is an oblique or slanting direction. A descending slope is a declivity (declination); and an ascending slope is an acclivity (inclination). For our purposes a slope is the rise (the vertical distance) over the run (the horizontal distance). Mathematically, the slope is the ratio of the change in the y-values (denoted by Δy, where Δ is the Greek letter delta) to the change in the x-values (denoted by Δx) for the coordinates of any two points on the line. The slope of the line parallel to the x-axis (horizontal line) is 0 since Δy is zero. A line parallel to the y-axis has no slope since $\frac{\Delta y}{\Delta x}$ is undefined for $\Delta x=0$ (division by zero is undefined). A line may have a positive or negative slope depending on the angle it forms with the x-axis. **Slopes of parallel lines** have the same slope. Vertical lines are parallel. Nonvertical lines are parallel if and only if they have the same slope and different y-intercepts. **Slopes of perpendicular** lines have slopes whose product is -1. Two lines with slopes m_1 and m_2 are perpendicular if and only if the product of their slopes is -1 ($m_1m_2=-1$). Lines are also perpendicular if one is vertical ($x=a$) and one is horizontal ($y=b$).

To find the slope m of a line the **slope formula** is used. The slope formula is the ratio of the distance, which is simply the difference of two points as shown below:

$$m=\frac{change\ in\ y}{change\ in\ x}=\frac{y_2-y_1}{x_2-x_1}=\text{rate of change}$$

Where m is the slope, (x_1, y_1) are the coordinates of a given point, and (x_2, y_2) represent the coordinates of a second point on the line. Figure 3 below shows the graph of the linear equation $y=\frac{2}{3}x+2$ with the slope and x and y-intercepts.

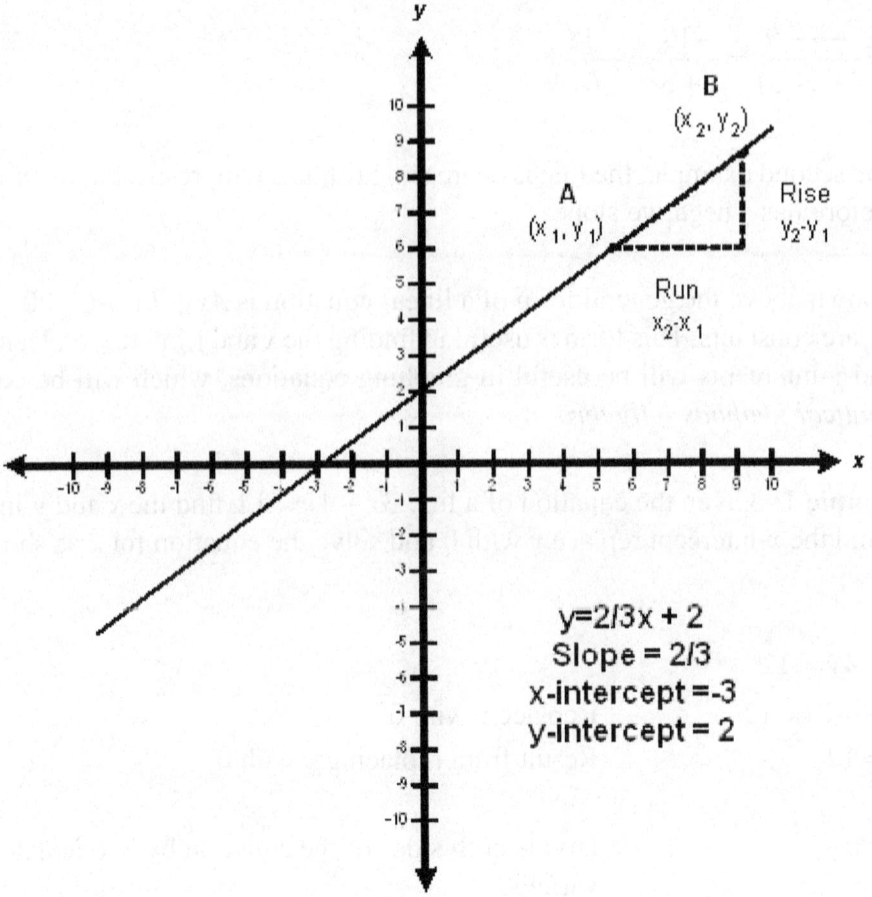

Figure 3. Graph of a linear equation

Example 1: Find the slope of the line that contains the points (-6, -2) and (-5, 1).

$$m = \frac{y_2 - y_1}{x_2 - x_1}$$

$$m = \frac{1-(-2)}{-5-(-6)} = \frac{1+2}{-5+6} = \frac{3}{1} = 3$$

In this example, $x_1 = -6, x_2 = -5, y_1 = -2$, and $y_2 = 1$, the line is increasing (acute with respect to the x-axis) and therefore the slope is positive.

Example 2: Find the slope of the line that contains the points (-5, 9) and (1, -9).

$$m = \frac{y_2 - y_1}{x_2 - x_1}$$

$$m = \frac{-9-9}{1-(-5)} = \frac{-18}{1+5} = \frac{-18}{6} = -3$$

In the second example, the line is decreasing (obtuse with respect to the x-axis) and therefore has a negative slope.

As shown above the general form of a linear equation is $Ax + By + C = 0$, where A, B, and C are constants. This form is useful in finding the x and y intercepts. Determining the x and y-intercepts will be useful in graphing equations, which will be covered in *Mathematical Methods - Algebra.*

Example 1: Given the equation of a line $3x - 4y = 12$ find the x and y-intercepts. To find the x-intercept replace y with 0 and solve the equation for x as shown below:

$3x - 4y = 12$	
$3x - 4(0) = 12$	Replace y with 0
$3x = 12$	Result from replacing y with 0
$\frac{3x}{3} = \frac{12}{3}$	Divide both sides of the equation by 3 to isolate the x variable
$x = 4$	4 is the x-intercept

To find the y-intercept replace x with 0 and solve the equation for y as shown below:

$3x - 4y = 12$	
$3(0) - 4y = 12$	Replace x with 0
$-4y = 12$	Result from replacing x with 0
$\frac{-4y}{-4} = \frac{12}{-4}$	Divide both sides of the equation by -4 to isolate the y variable
$y = -3$	-3 is the y-intercept

The **slope intercept form** of a linear equation as shown above is $y = mx + b$, where m is the slope and b is the y-intercept. To change a general form linear equation $Ax + By + C = 0$ to slope intercept form set the equation equal to y. For example:

$2x-3y-6=0$

$2x-3y+3y-6=0+3y$ Add $3y$ to both sides of the equation

$2x-6=3y$ Result of adding $3y$ to both sides of the equation

$\frac{2x}{3}-\frac{6}{3}=\frac{3y}{3}$ Divide both sides of the equation by 3 to isolate the y variable

$\frac{2}{3}x-2=y$ Result of dividing both sides of the equation by 3

$y=\frac{2}{3}x-2$ Write in standard form where the slope $m=\frac{2}{3}$ and the y-intercept $b=-2$

The **point slope form** of an equation of a line is $y-y_1=m(x-x_1)$ and is used to find the equation of a line when the slope and one coordinate of a line are known. This is a very important equation and is used extensively in calculus.

Example 1: A line has a slope of 2 and contains the point (-3, 1) find the equation of the line.

$y-y_1=m(x-x_1)$ Use the point slope form where $m=2$ and (x_1,y_1) equals (-3, 1)

$y-1=2(x-(-3))$

$y-1=2x+6$ Remove parentheses

$y-1+1=2x+6+1$ Add 1 to both sides of the equation

$y=2x+7$ This is the equation of the line

Example 2: Find the equation of a line if its slope is $\frac{3}{2}$ and the coordinates of a point on the line are (-4, -9).

$m=\frac{y_2-y_1}{x_2-x_1}$ Start with the slope formula

$\frac{3}{2}=\frac{y-(-9)}{x-(-4)}$ Insert the given coordinates (x_1,y_1)

$\frac{3}{2}=\frac{y+9}{x+4}$ Remove the parentheses. This is a proportion so you can cross multiply to simplify $3(x+4)=2(y+9)$

$3x+12=2y+18$	Result of cross multiplying is the point slope form
$3x-2y=6$	Group like terms - the variables on one side, the constants on the other side by subtracting 12 from both sides and subtracting $2y$ from both sides
$3x-3x-2y=6-3x$	Subtract $3x$ from both sides
$-2y=6-3x$	Result of subtracting $3x$ from both sides
$\frac{-2y}{-2}=\frac{6}{-2}-\frac{3x}{-2}$	Divide both sides of the equation by -2 to isolate the y variable
$y=\frac{3}{2}x-3$	Write in standard form

Example 3: Write an equation of the line that contains the points (6, 0) and (2, -6).

To find an equation of a line we need at least one point and the slope. Since the slope is not given we first need to find the slope of the line using the slope formula $m=\frac{y_2-y_1}{x_2-x_1}$. Once we have the slope we can use the point slope form $y-y_1=m(x-x_1)$ to find the equation of the line. If you forget the point slope form, you can always convert the slope formula by clearing the denominator and multiplying each side by $(x-x_1)$ a shown: $(x_2-x_1)m=\frac{y_2-y_1}{x_2-x_1}(x_2-x_1)\Rightarrow y-y_1=m(x-x_1)$.

Let $(y_2-y_1)=-6-0$ and $(x_2-x_1)=2-6$

$$m=\frac{-6-0}{2-6}=\frac{-6}{-4}=\frac{3}{2}$$

$y-0=\frac{3}{2}(x-6)$	Use the slope and the same coordinates $(x-x_1)$ in the point slope form
$y=\frac{3}{2}x-\frac{18}{2}$	Clear the parentheses
$y=\frac{3}{2}x-9$	Simplify - this is the equation of the line

Example 4: Given two linear equations $2y+4x=8$ and $5+2x=-y$ determine whether the lines are parallel.

We approach this type of problem as follows. We are given the fact that we have two linear equations. We can see that they are not in standard form $y=mx+b$. We are asked to determine whether the lines formed by the two equations are parallel. First, we must remember the definition of two lines that are parallel, which is that vertical lines are parallel, and nonvertical lines are parallel, if and only if they have the same slope and different *y*-intercepts. We need to determine if the equations represent a vertical or horizontal line or if the equations have the same slope. To do this we need to convert the equations to standard form as shown below.

$$2y+4x=8$$
$$2y+4x-4x=8-4x$$
$$2y=8-4x$$
$$\frac{2y}{2}=\frac{8}{2}-\frac{4x}{2}$$
$$y=-2x+4$$

$$5+2x=-y$$
$$5(-1)+2x(-1)=-y(-1)$$
$$y=-2x-5$$

We can see from the equation that the lines are not vertical ($x=a$) or horizontal ($y=b$). The slopes $m_1=-2$ and $m_2=-2$ are the same and the *y*-intercepts (0, 4) and (0, −5) are different, so the lines are parallel. In the equation $5+2x=-y$, we have multiplied the whole equation by -1 to remove the negative sign from the variable.

Example 5: Given two linear equations $y+2=5x$ and $5y+x=-15$ determine whether the lines are perpendicular.

We approach this problem as we did the previous problem. We are given the fact that we have two linear equations. We can see that they are not in standard form. We are asked to determine whether the lines formed by the two equations are perpendicular. First, we must remember the definition of two lines that are perpendicular. Two lines with slopes m_1 and m_2 are perpendicular if and only if the product of their slopes is -1 ($m_1m_2=-1$). Lines are also perpendicular if one is vertical ($x=a$) and one is horizontal ($y=b$).

Convert to standard form:

$$y+2=5x$$
$$y+2-2=5x-2$$

$$5y+x=-15$$
$$5y+x-x=-15-x$$

$y = 5x - 2$

$$5y = -15 - x$$

$$\frac{5y}{5} = \frac{-15}{5} - \frac{x}{5}$$

$$y = -3 - \frac{1}{5}x$$

$$y = -\frac{1}{5}x - 3$$

Next, find the slopes and determine if their product is equal to -1. After converting to standard form we can see that the lines are not horizontal or vertical and by definition $m_1m_2 = -1$. The slopes are $m_1 = 5$ and $m_2 = -\frac{1}{5}$, or $5\left(-\frac{1}{5}\right) = -1$, so the lines are perpendicular.

Summary

A linear equation is an equation that may be expressed in the form $ax + by = c$, where *a*, *b*, and *c* are any real numbers and $a \neq 0$. The main forms of linear equations are the general form of a linear equation: $Ax + By + C = 0$. The slope formula: $m = \frac{change\ in\ y}{change\ in\ x} = \frac{y_2 - y_1}{x_2 - x_1}$. The slope intercept form: $y = mx + b$, where *m* is the slope and *b* is the *y*-intercept, and the point-slope form (used to find the equation of a line): $y - y_1 = m(x - x_1)$.

The intercept of a straight line, curve, or surface on an axis of coordinates is the distance from the origin to the point where the line, curve, or surface cuts the axis. The *x*-intercept (*x*, 0) is the point at which the graph intersects the *x*-axis. It will always have a *y*-coordinate of zero. A horizontal line that is not the *x*-axis will not have an *x*-intercept. The *y*-intercept (0, *y*) is the point at which the graph intersects the *y*-axis. It will always have an *x*-coordinate of zero. A vertical line that is not the *y*-axis will not have a *y*-intercept.

The slope is the rise (the vertical distance) over the run (the horizontal distance). Mathematically, the slope is the ratio of the change in the *y*-values (denoted by Δy) to the change in the *x*-values (denoted by Δx) for the coordinates of any two points on the line. Slopes of parallel lines have the same slope. Vertical lines are parallel. Nonvertical lines are parallel if and only if they have the same slope and different *y*-intercepts. Slopes of perpendicular lines have slopes whose product is -1. Two lines with slopes m_1 and m_2 are perpendicular if and only if the product of their slopes is -1 ($m_1m_2 = -1$).

Ratio, Proportion, Percentage, and Variation

Ratio

Definition: A **ratio** is a comparison of two like numbers or quantities, and is the relative magnitude (size) of two quantities. For example, if the ratio between apples and oranges in a fruit basket is 3 to 2, then for every 3 apples there are 2 oranges. The ratio *a* to *b* can be written with either a colon *a*:*b* (read *a* to *b*) or with the fraction bar $\frac{a}{b}$. Ratios may also be expressed by the word per as in miles per hour. Per means for each, so rates are generally expressed as ratios reduced to a denominator of one. Since a ratio is the relationship between two quantities, the order of the relationship must be specific. For most ratio calculations when two terms are given, the first term given is the numerator and is called the antecedent (that which goes before) and the second term given is the denominator and is called the consequent (that which follows). For example, $\frac{a}{b}$ is read "*a* is to *b*". Whatever comes after the "to" goes second or is placed in the denominator.

The difference between ratios and fractions is that ratios represent part-per-part and fractions represent part-per-whole. Suppose a fruit basket has 6 apples and 4 oranges. The ratio of apples to oranges would be 6 to 4 or $\frac{6}{4}$ which is the same ratio as 3 to 2 or $\frac{3}{2}$. The fraction of apples and oranges in the basket is found by first determining the whole, or the total number of apples and oranges taken together. The total number of apples and oranges is $6+4=10$. So there is a total of 10 pieces of fruit. The fraction of apples is $\frac{6}{10}=\frac{3}{5}$ or 3 apples per every 5 pieces of fruit. The fraction of oranges is $\frac{4}{10}=\frac{2}{5}$ or 2 oranges per every 5 pieces of fruit.

Example 1: If the wing span of a plane is 76 ft. 6 inches, what will the wingspan of a scale model have to be if the dimensions of the plane and the scale model are 1:72?

$$\frac{1}{72}\times 76\frac{1}{2}=\frac{1}{72}\times\frac{153}{2}=\frac{153}{144}=1\frac{9}{144}=1\frac{1}{16}\text{ ft.}$$

Example 2: Three dealers receive 1,600 pounds of coffee in the ratio of 8:11:13. How many pounds should each dealer get?

To separate a quantity according to a given ratio, add the terms of the ratio to find the total number of parts. Find what fractional part each term is of the whole and divide the total quantity into parts corresponding to the fractional parts.

Add the terms of the ratio:

$8+11+13=32$	This becomes the denominator
$\frac{8}{32},\frac{11}{32},\frac{13}{32}$	Place the terms over the denominator. These are the fractional parts
$\frac{8}{32}\times 1600=400$	Multiply each fractional part times the total quantity of coffee to determine how much each dealer should get
$\frac{11}{32}\times 1600=550$	
$\frac{13}{32}\times 1600=650$	

Check: $400+550+650=1600$

Example 3: A baseball player had 5 hits in 10 at bats the first week of the season and 3 hits in 15 at bats the second week of the season. What was the average rate of hits per at bat for the two weeks?

The total number of hits is $5+3=8$, and the total number of at bats is $10+15=25$

The average batting rate is the total hits divided by the total at bats or $\frac{8}{25} = .320$

If you averaged the first week $\frac{5}{10} = .500$ and the second week $\frac{3}{15} = .200$ and then averaged these two rates you would get .350, which is incorrect.

An important fact to remember when solving rate problems is that the average rate is not equal to the average of the rates.

Proportion

Definition: A **proportion** is the relation between things (or parts of things) with respect to their comparative quantity, magnitude, or degree. Mathematically, a proportion is a statement of equality between two ratios. The four quantities that make up the ratios are called the terms of the proportion. In any proportion, the first and fourth terms are called the extremes and the second and third terms are called the means. Four numbers a, b, c, d, are in proportion when the ratio of the first pair equals the ratio of the second pair. This is denoted by $\frac{a}{b} = \frac{c}{d}$. The numbers a and d are the extremes, and the numbers b and c are the means.

A **direct proportion** exists when two quantities are related such that an increase in one quantity causes a corresponding increase in the other, or when a decrease in one quantity causes a corresponding decrease in the other. An **inverse proportion** exists when two quantities are related such that one is proportional to the reciprocal of the other and an increase in one quantity causes a corresponding decrease in the other, or when a decrease in one quantity causes a corresponding increase in the other. Typical inverse proportion problems include speed-and-time and work-and-time. For example, as the speed increases the time interval decreases, and as the number of men on the job increases the amount of time for the project decreases.

Algorithm: To solve a proportion, set the two ratios equal and cross multiply. Cross multiplying is most effective when the variable you are solving for is in the denominator of a fraction. If the variable you are solving for is in the numerator, it is easier to clear the denominator under the unknown you're solving for.

Example 1: Given $\frac{3}{7} = \frac{x}{63}$ find x by clearing the denominator.

$$\frac{3}{7} = \frac{x}{63}$$

$$63\left(\frac{3}{7}\right) = \left(\frac{x}{63}\right)63$$ Multiply both sides of the equation by 63

$$\frac{189}{7}=\frac{63}{63}x$$ Simplify, $\frac{63}{63}x$ is $1x$

$x=27$

Example 2: Given $\frac{3}{7}=\frac{x}{63}$ find x using cross multiplication.

$$\frac{3}{7}=\frac{x}{63}$$

$3(63)=7(x)$ Cross multiply

$189=7x$ Remove parentheses

$$\frac{189}{7}=\frac{7x}{7}$$ Divide both sides of the equation to isolate the variable x

$x=27$

Check: reduce $\frac{27}{63}$ or replace x with 27 and cross multiply $3(63)=189, 7(27)=189$

Example 3: A car travels 280 miles on 5 gallons of gas. What is the amount of gas required for a 700 mile trip?

$$\frac{280}{5}=\frac{700}{x}$$

$280x=700(5)$ Cross multiply

$280x=3500$

$$\frac{280x}{280}=\frac{3500}{280}$$ Divide both sides by 280 to isolate the x variable

$x=12.5$ gallons

Example 4: If 20 men build 8 cars per day, how many men are needed to build 12 cars per day?

The ratios here are 20 men 8 cars and x men 12 cars. The proportion would be set up as follows:

$$\frac{20}{8}=\frac{x}{12}$$

$20(12)=8x$ Cross multiply

$$240 = 8x$$

$$\frac{240}{8} = \frac{8x}{8}$$ Divide both sides by 8 to isolate the x variable

$$x = 30 \text{ men}$$

Percentage

Definition: A **percent** is a certain rate per one hundred or a part of a whole expressed in hundredths. For example, three percent means three out of every hundred or three hundredths of a given number. A percent is the term for expressing a fraction having a denominator of one hundred or in terms of an equivalent decimal. For example, 4% may be written as $\frac{4}{100}$ or 0.04. In computations, percent is called the rate, as in the rate of interest, denoted by the percent sign %. For example, 25% of a quantity is $\frac{25}{100} = \frac{1}{4} = .25$ of the whole. The **percentage** is the result obtained by multiplying a number by a percent (the percentage equals the rate times the base). In percentage problems, the rate (R) is the percent that is to be found, the base (B) is the whole quantity of which some percent is to be found, and the percentage (P) is the result obtained by taking a given percent of the base $P = BR$.

Algorithm: To work with percent problems you need to be able to convert a number with a percent sign to a decimal, and convert a number and fraction to a decimal. To convert a number written with a percent sign to a decimal drop the percent sign and move the decimal two places to the left. This is equivalent to dividing by 100. To convert a number and a fraction to a decimal, split the number and the fraction into two parts. For the whole number, move the decimal two places to the left as above. For the fractional part, divide the numerator by the denominator to put in decimal form. Add the decimal equivalent of the fractional part to the decimal equivalent of the whole number. Computations involving percentages generally involve three basic types of problems:

1. Finding the percentage, or the result obtained by taking a given percent (multiplying a number by a percent), or rate of a given quantity called the base (the percentage equals the rate times the base)
2. Finding the base when the percentage and rate are given
3. Finding the percent when the percentage and base are given.

For simple percentage problems the following formulas can be used:

$$\text{Part} = \text{Percent} \times \text{Whole}$$
$$\text{Percentage} = \text{Rate} \times \text{Base}$$

Although different words are used, as the two formulas above show, part and percentage are equivalent, as are percent and rate, and whole and base. The whole or base is generally associated with the word *of*, and the part or percentage is generally associated with the verbs *is/are*. To solve simple percentage problems, for *what* substitute the letter x, for *is* substitute an equal sign, and for *of* substitute a multiplication sign. The percent sign % translates to multiply by $\frac{1}{100}$ or .01.

Example 1: What is 12% of 25?

The percent (or rate) is 12% and the whole (or base) is 25. Therefore, $.12 \times 25 = 3$. So 3 is 12% of 25.

Set up as an equation you would have the following:

$$P = BR$$
$$P = 25(.12) = 3$$

This could also be translated as x (what) is (=) .12 of ($\times$) 25 (base)

Example 2: 45 is 3% of what number?

The part is 45 and the percent is 3. Therefore, using division which is the inverse of multiplication, $45 \div .03 = 1500$. So 45 is 3% of 1500. You can check this by taking 3% of 1500 ($1500 \times .03$) which is 45.

Set up as an equation you would have the following:

$$P = BR$$
$$45 = B(.03)$$
$$\frac{45}{.03} = \frac{B(.03)}{.03}$$
$$P = 1500$$

Example 3: 36 is what percent of 9?

The part is 36 and the whole is 9. Therefore, percent × 9 = 36. Using division which is the inverse of multiplication, $36 \div 9 = 4 = 400\%$

Set up as an equation you would have the following:

$P = BR$

$36 = B(9)$

$\frac{36}{9} = \frac{B(9)}{9}$

$B = 4$ or 400%

Example 4: What number increased by 25% of itself equals 120?

$B = \frac{P}{R}$ $\qquad$ $B = \frac{120}{1.25} = 96$

Due to the fact that 100% is 1, 100 percent of a number plus another 25 percent

$\frac{25}{100} = .25$, is 1.25.

Percent Change Problems

Two important concepts for understanding percentage problems, word problems, and statistics problems are difference and range. **Difference** is the quantity by which one quantity differs from another, or the remainder left after subtracting the one from the other. In mathematics, the difference is often the result, or answer, of a subtraction operation. The difference of two squares or two cubes is used for factoring in algebra, and the difference quotient is used in the definition of the derivative in calculus.

Generally, **range** means to change or differ within limits. The range of a variable is the set of values the variable takes on. In statistics, the range is the difference between the largest and the smallest observations in a sample, or the difference between the highest value and the lowest value.

There are two categories of percent change problems. There are amount and difference problems, and there are percent increase and percent decrease problems. The amount and difference problems are used when the rate and amount are given, and when the rate and difference are given respectively. The amount is equal to the base plus the percentage and the difference is equal to the base minus the percentage. The formulas for the amount and difference problems are as follows:

Amount formula: $\qquad base = \frac{amount}{1 + rate}$

Difference formula: $\qquad base = \frac{difference}{1 - rate}$

Example 1: The rent of an apartment is \$1,848 which is a 10% increase over the previous year. What was the rent the previous year?

$$base = \frac{1848}{1+.10} = \frac{1848}{1.10} = \$1,680$$

Example 2: A man sells his car for \$1,500 and loses 25% on the transaction. What was the initial cost of the car?

$$base = \frac{1500}{1-.25} = \frac{1500}{.75} = \$2,000$$

With percent increase and percent decrease problems there are several ways to describe the quantities. Essentially what you have is a ratio of the increase or decrease in a value, or the change in a value, divided by the original quantity or starting point value. The general form of the formula is as follows:

$$\frac{\text{Change}}{\text{Starting Point}} = \frac{\text{Increase or Decrease}}{\text{Original Quantity}} = \frac{\text{Difference}}{\text{Original Quantity}} \times 100 = \text{Percent Change}$$

Some examples will help clarify the use of this formula.

Example 3: A worker's salary increased from \$150 per week to \$200 per week. What is the percent change in the worker's salary?

$$\frac{\text{Difference}}{\text{Original Quantity}} \times 100 = \text{Percent Change}$$

$$\frac{200-150}{150} = \frac{50}{150} = .33(100) = 33\%$$

Example 4: A \$500 dress was on sale for \$400. What was the percent change in the price of the dress?

$$\frac{\text{Difference}}{\text{Original Quantity}} \times 100 = \text{Percent Change}$$

$$\frac{500-400}{500} = \frac{100}{500} = .20(100) = 20\%$$

Example 5: One month ago the price of apples was 20 percent less than it is now. If apples cost 90 cents a dozen, how much did they cost one month ago?

$$\frac{\text{Difference}}{\text{Original Quantity}} \times 100 = \text{Percent Change}$$

$\frac{\text{Difference}}{90} \times 100 = 20\%$ Multiply both sides by 90 and divide both sides by 100 to isolate the difference

$\text{Difference} = \frac{20}{100} \times 90$

$= .2 \times 90 = 18$ cents

Last month's price was 18 cents less than this month's. The price of apples was 72 cents one month ago. In this problem you have the original price (90 cents) and the percent change (20%) so you need to find the difference (18 cents). Once you find the difference you subtract the difference from the current price to find the old price $90 - 18 = 72$.

Variation

Definition: To vary is to exhibit or undergo change or to take on successive values. Variation is the act or process of varying, or the amount or rate of change. Mathematically, **variation** is the relationship between two quantities in which a change in one quantity results in a change in the other quantity. A quantity which never changes in value is called a **constant of variation** or variation constant. For example, in the formula $C = 2\pi r$ where C is the circumference and r is the radius, 2 and π are constants. There are four types of variation, **direct**, **inverse, joint**, and **combined** variation.

Variation should not be confused with variance, which is used in statistics. In statistics, variance is a measure of dispersion (spread) for a list of numbers. Dispersion is the degree to which a group of numbers are scattered away for their average.

Direct variation is when two variables are so related that their ratio remains constant and one of the variables is said to vary directly or vary proportionately as the other. For example, when velocity is constant, the distance s traveled per unit time is proportional to the time t or $s = kt$, where k is the constant of variation. Direct variation can be described in several different ways. For example, for a linear function $f(x) = kx$ or equivalently $y = kx$ where k is a positive constant, there exists a direct variation, or y varies directly as x, or y is directly proportional to x. The number k is called the variation constant, or constant of proportionality.

Example 1: The circumference (the length of the circle) C of a circle varies proportionally to its diameter d. If the circumference is 25 inches what is the diameter? The formula for the circumference of a circle is as follows:

$C = \pi d, or\ C = \pi r^2$

Where r in the second formula is the radius squared which is equal to the diameter. In this formula, π is the constant of variation. Since we are given C, and we know the value of π is approximately 3.1415 we can solve the equation for the unknown, the diameter d.

$$C = \pi d$$
$$25 = 3.1415d$$

$$\frac{25}{3.1415} = \frac{3.1415d}{3.1415}$$

Divide both sides by the constant 3.1415 to isolate the variable d

$$d = 7.95 \text{ inches}$$

Inverse variation is when the ratio of one variable to the reciprocal of the other is constant. For a linear function $f(x) = \frac{k}{x}$ or equivalently $y = \frac{k}{x}$ where k is a positive constant, there exists an inverse variation, or y varies inversely as x, or y is inversely proportional to x. With inverse variation, as x gets larger, y gets smaller. As with direct variation, k is called the variation constant or constant of proportionality.

Example: The intensity of light varies inversely with the square of the distance d from the light source. If the intensity of the light source is 80 lumens at a distance of 20 feet, what is the intensity at 40 feet?

The formula for intensity I is as follows:

$$I = \frac{k}{d^2}$$

Where I is the intensity, d is the distance and k is the variation constant with a value of 32,000

$$I = \frac{32000}{40^2} = \frac{32000}{1600} = 20 \text{ lumens}$$

The intensity decreases as the distance increases.

Joint variation is when one variable varies directly as the product of two variables, the one variable is said to vary jointly as the other two. For example, when $x = kyz$, x varies jointly as y and z.

Example: The volume of wood V in a tree varies jointly as the height h and the square of the circumference c. If the height of the tree is 75 meters, the circumference is 2 meters, and the constant of variation is 3.2, what is the volume of wood?

The formula for the volume V of wood is as follows:

$V = khc^2$ Where V is the volume, k is the constant of variation which in this case is 3.2, h is the height, and c is the circumference.

$V = 3.2(75)(2)^2 = 960$ cubic meters (m^3)

Combined variation is when one quantity varies as some combination of other quantities, such as z varying directly as y and inversely as x. For example, in the equation

$y = k\frac{xz}{w^2}$, y varies jointly as x and z (the product of two variables) and inversely as the square of w.

Example: The resistance of any round conductor varies jointly as the length and inversely as the square of the diameter. What is the electrical resistance of 1,000 feet of copper wire .01 inch in diameter, with $k = 10.3$?

The formula for resistance R is as follows:

$R = \frac{kL}{d^2}$ Where R is the resistance in ohms, L is the length, d is the diameter in mils, and k is equal to 10.3

$$R = \frac{10.3(1000)}{(100)^2} = 1.03 \text{ ohms}$$

Summary

A ratio is a comparison of two like numbers or quantities, and is the relative magnitude (size) of two quantities. Since a ratio is the relationship between two quantities, the order of the relationship must be specific. The difference between ratios and fractions is that ratios represent part-per-part and fractions represent part-per-whole.

A proportion is the relation between things (or parts of things) with respect to their comparative quantity, magnitude, or degree. Mathematically, a proportion is a statement of equality between two ratios. A direct proportion exists when two quantities are related such that an increase in one quantity causes a corresponding increase in the other, or when a decrease in one quantity causes a corresponding decrease in the other. An inverse proportion exists when two quantities are related such that one is proportional to the reciprocal of the other and an increase in one quantity causes a corresponding decrease in the other, or when a decrease in one quantity causes a corresponding increase in the other.

A percent is a certain rate per one hundred or a part of a whole expressed in hundredths. The percentage is the result obtained by multiplying a number by a percent (the percentage equals the rate times the base). In percentage problems the rate (R) is the percent that is to be found, the base (B) is the whole quantity of which some percent is to be found, and the percentage (P) is the result obtained by taking a given percent of the base $P = BR$.

Variation is the relationship between two quantities in which a change in one quantity results in a change in the other quantity. There are four types of variation, direct, inverse, joint, and combined variation. Direct variation is when two variables are so related that their ratio remains constant and one of the variables is said to vary directly or vary proportionately as the other. Inverse variation is when the ratio of one variable to the reciprocal of the other is constant. Joint variation is when one variable varies directly as the product of two variables, the one variable is said to vary jointly as the other two. Combined variation is when one quantity varies as some combination of other quantities, such as z varying directly as y and inversely as x.

Sequences, Summation, and Series

Sequences

Definition: A **sequence** is a set of numbers in which each one is related in a definite way to the number that precedes it. There are two types of sequences, the arithmetic sequence and the geometric sequence.

Arithmetic Sequence

Definition: An **arithmetic sequence** is a number series in which each term may be obtained from the preceding one by adding a constant called the common difference. If an arithmetic sequence has first term a_1 and common difference *d*, then a_n the *n*th term of the sequence, is given by: $a_n = a_1 + (n-1)d$.

Example: Find the 25th term of the arithmetic sequence 4 + 7 + 10 + 13 + … Since $a_1 = 4$, $d = 3$, and $n = 25$:

$$a_n = a_1 + (n-1)d$$
$$a_n = 4 + (25-1)3 = 76$$

The sum S_n of the first *n* terms of an arithmetic sequence is given by:

$$S_n = \frac{n}{2}\left[2a_1 + (n-1)d\right] \text{ Or equivalently, } S_n = \frac{n}{2}(a_1 + a_n)$$

Example: Find the sum of the first 6 terms of 4 + 7 + 10 + ...

$$S_n = \frac{n}{2}\left[2a_1 + (n-1)d\right]$$

$$S_n = \frac{6}{2}\left[2(4) + (6-1)3\right] = 3(8+15) = 69$$

Geometric Sequence

Definition: A **geometric sequence** is a number series in which each term may be obtained from the preceding one by multiplying by a fixed number called the common ratio. If a geometric sequence has a first term a_1 and a common ratio r, then the nth term of a geometric sequence is given by: $a_n = a_1 r^{n-1}$.

Example 3: Find the 8th term of the geometric sequence 3, 6, 12, 24, ... Since the first term is $a_1 = 3$, the ratio is $r = 2$, and $n - 1 = 8 - 1 = 7$.

$$a_n = a_1 r^{n-1}$$

$$a_n = 3(2^7) = 3(128) = 384$$

If a geometric sequence has a first term a_1 and a common ratio r, then the sum S_n of the first n terms is given by:

$$S_n = \frac{a_1(r^n - 1)}{r - 1}, \quad r \neq 1$$

Example 4: Find the sum of the first 5 terms of 2, 16, 18, ..., Since $a = 2$, $n = 5$, and $r = 3$:

$$S_n = \frac{2(3^5 - 1)}{3 - 1} = \frac{2(243 - 1)}{3 - 1} = \frac{2(242)}{2} = 242$$

The **common ratio** is the number by which each term in a geometric progression is multiplied to obtain the term immediately following it. The common ratio is determined as follows:

$$r = \frac{a_n}{a_{n-1}}$$

Example 5: Find the common ratio for the sequence 3, 9, 27, 81, …

To find the common ratio divide any term (other than the first term) by the preceding term. In this example, the third and fourth terms will be used.

$$r=\frac{81}{27}=3$$

Summation

Definition: Summation is the act or process of forming a sum.

Sigma Notation

The Greek letter sigma (Σ) is used for summation notation. The sum of *n* terms $a_1, a_2, a_3, \ldots a_n$ is written as

$$\sum_{i=1}^{n} a_i = a_1 + a_2 + a_3 + \ldots + a_n$$

Where *i* is the index of summation, a_i is the *i*th term of the sum, and the upper and lower bounds of summation are *n* and 1.

For example, the first four terms of the sequence 3, 5, 7, 9,…, 2*k*+1 can be written as follows:

$$\sum_{k=1}^{4} 2k+1$$

This is read "the sum as *k* goes from 1 to 4 of 2*k*+1.

Summation Rules

1. $\sum_{i=1}^{n} ca_i = c\sum_{i=1}^{n} a_i$

2. $\sum_{i=1}^{n} (a_i \pm b_i) = \sum_{i=1}^{n} a_i \pm \sum_{i=1}^{n} b_i$

3. $\sum_{i=1}^{n} c = c + c + c + \ldots + c$ Equals *c, n* times (*cn*), where *c* is a constant.

4. $\sum_{i=1}^{n} i = 1+2+3+...+n = \frac{n(n+1)}{2}$

5. $\sum_{i=1}^{n} i^2 = 1^2+2^2+3^2+...+n^2 = \frac{n(n+1)(2n+1)}{6}$

6. $\sum_{i=1}^{n} i^3 = 1^3+2^3+3^3+...+n^3 = \frac{n^2(n+1)^2}{4}$

7. $\sum_{i=1}^{n} i^4 = \frac{n(n+1)(2n+1)(3n^2+3n-1)}{30}$

Series

Definition: A **series** is an indefinite number of terms succeeding one another, each of which is derived from one or more of the preceding by a fixed law, called the law of the series; as, an arithmetical series; a geometrical series. A series is the indicated sum of a finite or infinite sequence of terms. If the sequence of partial sums converges to a definite value, the series is said to converge. If the sequence of partial sums does not converge to a limit (e.g., it oscillates or approaches $\pm\infty$), the series is said to diverge. Limits are critical to understanding calculus and will be covered in detail in *Mathematical Methods - Calculus*. There are two types of series, convergent and divergent.

Convergent Series

Definition: A **convergent series** is a series whose sequence of partial sums approaches a limit. To converge is to approach nearer together or tend to one point. The series $\sum_{n=0}^{\infty}\left(\frac{1}{2}\right)^n$ converges since its sequence of partial sums is the sequence $1, 1\frac{1}{2}, 1\frac{3}{4}, 1\frac{7}{8}, ...,$ whose limit is 2.

Divergent Series

Definition: A **divergent series** is a series whose sequence of partial sums does not approach a limit. To diverge is to tend to spread apart or to extend from a common point in different directions. The series $\sum_{n=1}^{\infty}\frac{1}{n}$ does not converge since its sequence of partial sums gets larger without limit.

Summary

A sequence is a set of numbers in which each one is related in a definite way to the number that precedes it. An arithmetic sequence is a number series in which each term may be obtained from the preceding one by adding a constant called the common difference. Summation is the act or process of forming a sum. The Greek letter sigma (Σ) is used for summation notation.

A series is an indefinite number of terms succeeding one another, each of which is derived from one or more of the preceding by a fixed law, called the law of the series; as, an arithmetical series; a geometrical series. A series is the indicated sum of a finite or infinite sequence of terms. If the sequence of partial sums converges to a definite value, the series is said to converge. If the sequence of partial sums does not converge to a limit (e.g., it oscillates or approaches $\pm\infty$), the series is said to diverge.

A convergent series is a series whose sequence of partial sums approaches a limit. A divergent series is a series whose sequence of partial sums does not approach a limit. Series and limits are important concepts in calculus.

Appendix A

The Mathematics Subject Classification (MSC) Scheme

This list is for the major headings only. The full list with subcategories is available at the American Mathematical Society (www.ams.org).

00: General material including elementary mathematics
01: History and biography
03: Mathematical logic and foundations
05: Combinatorics and graph theory
06: Order, lattices, ordered algebraic structures
08: General algebraic systems
11: Number theory
12: Field theory and polynomials
13: Commutative rings and algebras
14: Algebraic geometry
15: Linear and multilinear algebra; matrix theory
16: Associative rings and algebras
17: Nonassociative rings and algebras
18: Category theory, homological algebra
19: K-theory
20: Group theory and generalizations
22: Topological groups, Lie groups
26: Real functions and elementary calculus
28: Measure and integration
30: Functions of a complex variable
31: Potential theory
32: Several complex variables and analytic spaces
33: Special functions including trigonometric functions

34: Ordinary differential equations
35: Partial differential equations
37: Dynamical systems and ergodic theory
39: Difference and functional equations
40: Sequences, series, summability
41: Approximations and expansions
42: Fourier analysis
43: Abstract harmonic analysis
44: Integral transforms operational calculus
45: Integral equations
46: Functional analysis
47: Operator theory
49: Calculus of variations and optimal control; optimization
51: Geometry, including classic Euclidean geometry
52: Convex and discrete geometry
53: Differential geometry
54: General topology
55: Algebraic topology
57: Manifolds and cell complexes
58: Global analysis, analysis on manifolds
60: Probability theory and stochastic processes
62: Statistics
65: Numerical analysis
68: Computer science
70: Mechanics of particles and systems
74: Mechanics of deformable solids
76: Fluid mechanics
78: Optics, electromagnetic theory
80: Classical thermodynamics, heat transfer
81: Quantum Theory
82: Statistical mechanics, structure of matter
83: Relativity and gravitational theory
85: Astronomy and astrophysics
86: Geophysics
90: Operations research, mathematical programming
91: Game theory, economics, social and behavioral sciences
92: Biology and other natural sciences
93: Systems theory; control
94: Information and communication, circuits
97: Mathematics education

Appendix B

Arithmetic Tables

To use the tables, select a number from the first column (vertical) of the table, then select the number you wish to add, subtract, multiply or divide from the top row (horizontal) of the table. The first column and row are in **bold** type. The answer will be where the column and row intersect. For example, to add 7 to 8, find 7 in the first column then move right in the same row until you are under the 8 column. The sum is 15 as highlighted below. For addition and multiplication you can start with the top row and select a number from the first column and you will get the same answer. This will not work for subtraction and division because the order of the numbers determines the value (they are not commutative). See properties of real numbers in Appendix C.

Addition Table

	1	**2**	**3**	**4**	**5**	**6**	**7**	**8**	**9**	**10**	**11**	**12**
1	2	3	4	5	6	7	8	9	10	11	12	13
2	3	4	5	6	7	8	9	10	11	12	13	14
3	4	5	6	7	8	9	10	11	12	13	14	15
4	5	6	7	8	9	10	11	12	13	14	15	16
5	6	7	8	9	10	11	12	13	14	15	16	17
6	7	8	9	10	11	12	13	14	15	16	17	18
7	8	9	10	11	12	13	14	**15**	16	17	18	19
8	9	10	11	12	13	14	15	16	17	18	19	20
9	10	11	12	13	14	15	16	17	18	19	20	21
10	11	12	13	14	15	16	17	18	19	20	21	22
11	12	13	14	15	16	17	18	19	20	21	22	23
12	13	14	15	16	17	18	19	20	21	22	23	24

Subtraction Table

	1	**2**	**3**	**4**	**5**	**6**	**7**	**8**	**9**	**10**	**11**	**12**
1	0	-1	-2	-3	-4	-5	-6	-7	-8	-9	-10	-11
2	1	0	-1	-2	-3	-4	-5	-6	-7	-8	-9	-10
3	2	1	0	-1	-2	-3	-4	-5	-6	-7	-8	-9
4	3	2	1	0	-1	-2	-3	-4	-5	-6	-7	-8
5	4	3	2	1	0	-1	-2	-3	-4	-5	-6	-7
6	5	4	3	2	1	0	-1	-2	-3	-4	-5	-6
7	6	5	4	3	2	1	0	-1	-2	-3	-4	-5
8	7	6	5	4	3	2	1	0	-1	-2	-3	-4
9	8	7	6	5	4	3	2	1	0	-1	-2	-3
10	9	8	7	6	5	4	3	2	1	0	-1	-2
11	10	9	8	7	6	5	4	3	2	1	0	-1
12	11	10	9	8	7	6	5	4	3	2	1	0

Multiplication Table												
	1	**2**	**3**	**4**	**5**	**6**	**7**	**8**	**9**	**10**	**11**	**12**
1	1	2	3	4	5	6	7	8	9	10	11	12
2	2	4	6	8	10	12	14	16	18	20	22	24
3	3	6	9	12	15	18	21	24	27	30	33	36
4	4	8	12	16	20	24	28	32	36	40	44	48
5	5	10	15	20	25	30	35	40	45	50	55	60
6	6	12	18	24	30	36	42	48	54	60	66	72
7	7	14	21	28	35	42	49	56	63	70	77	84
8	8	16	24	32	40	48	56	64	72	80	88	96
9	9	18	27	36	45	54	63	72	81	90	99	108
10	10	20	30	40	50	60	70	80	90	100	110	120
11	11	22	33	44	55	66	77	88	99	110	121	132
12	12	24	36	48	60	72	84	96	108	120	132	144

Division Table												
	1	**2**	**3**	**4**	**5**	**6**	**7**	**8**	**9**	**10**	**11**	**12**
1	1.000	0.500	0.333	0.250	0.200	0.167	0.143	0.125	0.111	0.100	0.091	0.083
2	2.000	1.000	0.667	0.500	0.400	0.333	0.286	0.250	0.222	0.200	0.182	0.167
3	3.000	1.500	1.000	0.750	0.600	0.500	0.429	0.375	0.333	0.300	0.273	0.250
4	4.000	2.000	1.333	1.000	0.800	0.667	0.571	0.500	0.444	0.400	0.364	0.333
5	5.000	2.500	1.667	1.250	1.000	0.833	0.714	0.625	0.556	0.500	0.455	0.417
6	6.000	3.000	2.000	1.500	1.200	1.000	0.857	0.750	0.667	0.600	0.545	0.500
7	7.000	3.500	2.333	1.750	1.400	1.167	1.000	0.875	0.778	0.700	0.636	0.583
8	8.000	4.000	2.667	2.000	1.600	1.333	1.143	1.000	0.889	0.800	0.727	0.667
9	9.000	4.500	3.000	2.250	1.800	1.500	1.286	1.125	1.000	0.900	0.818	0.750
10	10.000	5.000	3.333	2.500	2.000	1.667	1.429	1.250	1.111	1.000	0.909	0.833
11	11.000	5.500	3.667	2.750	2.200	1.833	1.571	1.375	1.222	1.100	1.000	0.917
12	12.000	6.000	4.000	3.000	2.400	2.000	1.714	1.500	1.333	1.200	1.091	1.000

Appendix C

Properties of Real Numbers

For any real numbers *a*, *b*, and *c*:

$a+b=b+a$ $ab=ba$	Commutative properties of addition and multiplication
$a+(b+c)=(a+b)+c$ $a(bc)=(ab)c$	Associative properties of addition and multiplication
$a+0=0+a=a$	Additive identity property
$-a+a=a+(-a)=0$	Additive inverse property
$a\cdot 1=1\cdot a=a$	Multiplicative identity property
$a\cdot\frac{1}{a}=\frac{1}{a}\cdot\frac{a}{1}=1,\ a\neq 0$	Multiplicative inverse property
$ab\pm ac=a(b\pm c)$	Distributive property

Associative property (grouping) - A property of real numbers that states the sum or product of a set of numbers or variables has the same value, regardless of how the numbers or variables are grouped.

Commutative property (order) - A property of real numbers that states the sum or product of two terms is unaffected by the order in which the terms are added or multiplied. For example; the sum or product remains the same; independent of order; as in $a+b=b+a$. Subtraction and division are not commutative for the set of real numbers. For example:

$$6-3=3\neq 3-6=-3 \text{ and } 6\div 3=2\neq 3\div 6=\frac{1}{2}$$

Distributive property (separation) - A property of real numbers that states the product of a number and the sum or difference of two numbers is the same as the sum or difference of their products.

Appendix D

Translating English Terms into Algebraic Symbols

English Term	Algebraic Symbols
A number *increased* by six	$n+6$
Five *more* than a number	$x+5$
Two *less* than a number	$x-2$
Four *times* as much	$4x$
The *product* of a number and three	$3x$
Twice a number *decreased* by six	$2n-6$
Four *less* than five *times* a number	$5n-4$
The *sum* of three *times* a number *and* twelve	$3n+12$
The *quotient* of a number and eight	$\frac{x}{8}$
One half a number is fifty	$\frac{1}{2}n=50$
If from *twice* a number you *subtract* four the difference *is* twenty	$2n-4=20$
Three *is* four *more* than a number	$3=x+4$
Twenty five *is* nine *more* than four *times* a number	$25=9+4n$
Sixteen *subtracted* from five *times* a number *equals* the number *plus* four	$5n-16=n+4$
Twenty five *is* the same as ten *added* to *twice* a number	$25=10+2n$
Three *x* *is* five *less* than *twice* *x*	$3x=2x-5$
The *sum* of two *y* and three *is* the same as the *difference* of three *y* and one	$2y+3=3y-1$
Twice the *quantity of* two *y and* six	$2(2y+6)$
Twice the *quantity* of seven *plus x is* the same as the *difference* of *x* and seven	$2(7+x)=x-7$
The *sum* of two *y* and the *quantity* of three *plus y plus twice* the *quantity* two *y minus* two *is* fifteen	$2y+(3+y)+2(2y-2)=15$
Suzie's age (x) four years from now	$x+4$
Sarah's age (x) ten years ago	$x-10$
Number of cents in x quarters	$25x$
Number of cents in $2x$ dimes	$10(2x)$
Number of cents in $x + 5$ nickels	$5(x+5)$

Separate seventeen into two parts	x and $17-x$
Distance traveled in x hours at fifty miles per hour	$50x$
Two consecutive integers	x and $x+1$
Two consecutive even integers	x and $x+2$
Two consecutive odd integers	x and $x+2$
$10,000 *separated* into two investments	x and $10,000-x$

General Terms for Solving Word Problems

1. **Times as much** means multiply
2. **More than** means add
3. **Decreased** means subtract
4. **Increased** means add
5. **Percent of** means multiply
6. **Is, was, will be**, means equal to
7. **Separate** 28 into to parts means find two numbers whose sum is 28
8. If 7 exceeds 2 by 5 then 7 – 2 = 5. **Exceeds** becomes a minus sign (-) and **by** becomes an equals sign (=).
9. Nine less than twelve means 12 – 9
10. Six less two means 6 - 2
11. Four less than three times y means $3y$-4
12. Four times the quantity six plus two means 4(6 + 2)
13. One half a number times fifteen means $\frac{1}{2}n(15)$
14. One third a number less two means $\frac{1}{3}n-2$
15. Ten times the sum of twice a number and six means $10(2n+6)$
16. Ten times the difference of a number and ten $10(n-10)$
17. Eight times the quantity y plus two divided by four means $\frac{8(y+2)}{4}$
18. What part of 11 is 8 means $x(11)=8$
19. 160 is $\frac{4}{5}$ of what number means $160=\frac{4}{5}x$

Appendix E

Solving Word Problems

Steps for solving word problems:

1. Read the problem all the way through to determine what type of problem it is and to determine what information is given.
2. Look for a question at the end of the problem to determine what you are solving for. Sometimes there is more than one solution.
3. Start every problem by letting x = something.
4. If you have to find more than one quantity or unknown try to determine the smallest unknown. This unknown is often the one to set equal to x.
5. Translate the words to symbols and solve the resulting equation.
6. Check your work.

Number Problem Example 1: The sum of three consecutive integers is 54. Find the integers.

Let x = the first consecutive integer
$x + 1$ = the second consecutive integer
$x + 2$ = the third consecutive integer
54 = the sum of three consecutive integers

$$x+(x+1)+(x+2)=54$$
$$3x+3=54$$
$$3x+3-3=54-3$$

$$\frac{3x}{3}=\frac{51}{3}$$

$$x=17$$
$$x+1=18$$
$$x+2=19$$

Check: $17+18+19=54$

Number Problem Example 2: There were 10,483 people who attended a rock concert. If there were 811 more boys than girls, and 2481 fewer adults over 50 years of age than there were girls, how many of each group attended the concert?

Let: $x =$ the number of girls
$x + 811 =$ the number of boys
$x - 2481 =$ the number of adults

$$x + x + 811 + x - 2481$$
$$3x - 1670 = 10483$$
$$3x - 1670 + 1670 = 10483 + 1670$$
$$3x = 12153$$

$$\frac{3x}{3} = \frac{12153}{3}$$

$x = 4051$ Number of girls
$x + 811 = 4862$ Number of boys
$x - 2481 = 1570$ Number of adults

Mixture Problem Example 1: A store manager wishes to reduce the price on her fresh ground coffee by mixing two grades. If she has 50 pounds of coffee which sells for \$1.20 per pound, how much coffee worth 80 cents per pound must she mix with it so she can sell the final mixture for \$1.00 per pound?

Let x = pounds of 80 cent coffee

$50 \times 1.20 =$ Total cost for the \$1.20 per pound coffee
$0.80x =$ Total cost of the 80 cent per pound coffee
$1(x + 50) =$ Pounds of 80 cent coffee

$50(1.20) + .80x = 1(x + 50)$ Clear the decimal by multiplying by 100
$6000 + 80x = 100x + 5000$ Subtract 6000 from each side of the equation

$80x = 100x - 1000$
$80x - 100x = 100x - 1000 - 100x$ Subtract 100x from both sides of the equation

$$-20x = -1000$$

$$\frac{-20x}{-20} = \frac{-1000}{-20}$$

$x = 50$ 50 pounds of 80 cent coffee

Mixture Problem Example 2: A doctor orders 20 grams of a 52% solution of medicine. The pharmacist has only bottles of 40% and 70% solution. How much of each must he use to obtain the 20 grams of the 52% solution?

Let: x = grams of 40% solution
$20 - x$ = grams of 70% solution

$.4x + .70(20 - x) = .52(20)$ Clear parentheses
$.4x + 14 - .7x = 10.4$ Clear decimals by multiplying by 10
$4x + 140 - 7x = 104$ Combine like terms (4*x* - 7*x*)
$140 - 3x = 104$ Subtract 140 from each side of the equation

$140 - 140 - 3x = 104 - 140$
$-3x = -36$

$\frac{-3x}{-3} = \frac{-36}{-3}$ Divide each side by -3 to isolate the *x* variable

$x = 12$ Grams of the 40% solution
$20 - x = 8$ Grams of the 70% solution

Coin Problem Example: A collection of coins has a value of 64 cents. There are two more nickels than dimes and three times as many pennies as dimes. How many of each kind of coin are there?

Number of Coins	**Value in Cents**
$x =$ number of dimes	$10x =$ value of dimes
$x + 2 =$ number of nickels	$5(x + 2) =$ value of nickels
$3x =$ number of pennies	$3x =$ value of pennies

$10x + 5(x + 2) + 3x = 64$ Set number of coins equal to 64
$10x + 5x + 10 + 3x = 64$ Remove parentheses
$18x + 10 = 64$ Combine like terms (10*x* + 5*x* + 3*x*)
$18x + 10 - 10 = 64 - 10$ Subtract 10 from each side

$\frac{18x}{18} = \frac{54}{18}$ Divide both sides by 18 to isolate the *x* variable

$x = 3$
$x + 2 = 5$
$3x = 9$

Check: $3(10)+5(5)+9(1)=64$

Age Problem Example: A man is 7 times as old as his son. In two years the father will be only 5 times as old as his son. What is the age of each?

$x=$ son's age now
7x = father's age now
$x+2$ = son's age in 2 years
$7x+2=$ father's age in 2 years

$7x+2=5(x+2)$	Set father's age equal to 5 times son's age
$7x+2=5x+10$	Clear parentheses
$7x-5x=10-2$	Subtract 2 and 5*x* from both sides
$2x=8$	
$\frac{2x}{2}=\frac{8}{2}$	Divide both sides by 2 to isolate the *x*

$x=4$, son' age now
$7x=(7\times4)=28$, father's age now

Lever Problem Example: A 200 pound man is sitting on one end of a lever, 12 feet from the fulcrum. What weight resting on the opposite end and 8 feet from the fulcrum would bring the lever into balance?

The weight times its distance from the fulcrum on one side equals the weight times its distance from the fulcrum on the other side. This can be represented by the formula $w_1d_1=w_2d_2$ where w_1 and d_1 represent one weight and distance and w_2 and d_2 represent the other.

$3x=12(200)$
$3x=2400$

$$\frac{3x}{3}=\frac{2400}{3}$$
$x=800$

Finance Problem Example 1: James inherited two different stocks whose yearly income was \$2100. The total appraised value of the stocks was \$40,000 and one was paying a 4% dividend yield and the other was paying a 6% dividend yield. What was the value of each stock?

Let: $x =$ Value in dollars of the stock paying 4%

$40000 - x =$ Value in dollars of the stock paying 6%

$.04x =$ Interest on stock paying 4%

$.06(40000 - x) =$ Interest on stock paying 6%

$.04x + .06(40000 - x) = 2100$	Remove parentheses
$.04x + 2400 - .06x = 2100$	Multiply by 100 to clear decimals
$240000 - 2x = 210000$	Combine like terms
$-2x = -30000$	Subtract 240000 from both sides of the equation
$\frac{-2x}{-2} = \frac{-30000}{-2}$	Divide both sides by -2
$x = 15,000$	Value of stock paying 4%
$40000 - x = 25,000$	Value of stock paying 6%

Check:

$$.04(15000) + .06(40000 - 15000) = 2100$$
$$600 + 2400 - 900 = 2100$$
$$2100 = 2100$$

Finance Problem Example 2: The total of two investments is \$2500. One amount is invested at 5% and the other at 6%. The annual interest from the latter is \$62 more than the former. How much is invested at each rate?

Let: $x =$ Amount in dollars invested at 5%

$2500 - x =$ Amount in dollars invested at 6%

$.05x =$ Yearly interest on 5% investment

$.06(2500 - x) =$ Yearly interest on 6% investment

$.06(2500 - x) = .05x + 62$	
$150 - .06x = .05x + 62$	Clear parentheses
$1500 - 6x = 5x + 6200$	Multiply the whole equation by 100 to clear decimals

$1500 - 6x - 5x = 5x - 5x + 6200$ Subtract 5*x* from both sides of the equation

$1500 - 11x = 6200$

$1500 - 1500 - 11x = 6200 - 1500$ Subtract 1500 from both sides of the equation

$-11x = -8800$

$$\frac{-11x}{-11} = \frac{-8800}{-11}$$

$x = 800$ Amount invested at 5%

$2500 - x = 1700$ Amount invested at 6%

Check:

$$.06(2500 - 800) = .05(800) + 62$$
$$150 - 48 = 40 + 62$$
$$102 = 102$$

Rate and Distance Problem Example 1: Two planes leave St. Louis at 1:00 PM. Plane A heads east at 450 miles per hour and Plane B head due west at 600 miles per hour. How long will it be before the planes are 2100 miles apart?

The distance is equal to rate times time $r \times t = d$. If you know two of the variables you can find the third. To set up the equations for these types of problems it is helpful to set up a table of variables.

	Rate	**Time**	**Distance**
Plane A	450	x	$450x$
Plane B	600	x	$600x$

$450x + 600x = 2100$

$1050x = 2100$ Combine like terms

$$\frac{1050x}{1050} = \frac{2100}{1050}$$ Divide both sides by 1050 to isolate the variable

$x = 2$ hours

Rate and Distance Problem Example 2: Sarah flies to San Francisco from Los Angeles in 4 hours. She flies back in 3 hours. If the wind is blowing from the north at a velocity of 20 miles per hour during both flights, what was the airspeed of the plane?

In this type of problem, the plane's speed in still air (the planes airspeed) would be increased by a tailwind and decreased by a headwind.

Let: $x =$ the airspeed of the plane
$20 =$ the velocity of the wind
$x - 20 =$ ground speed going toward San Francisco against the wind
$x + 20 =$ ground speed going toward Los Angeles with a tailwind

	Rate	**Time**	**Distance**
LA to SF	$x-20$	4	$4(x-20)$
SF to LA	$x+20$	3	$3(x+20)$

The distances are equal.

$4(x-20) = 3(x+20)$

$4x - 80 = 3x + 60$ Remove parentheses

$4x - 3x - 80 = 3x - 3x + 60$ Subtract $3x$ from both sides of the equation

$x - 80 = 60$

$x - 80 + 80 = 60 + 80$ Add 80 to both sides of the equation to isolate the x variable

$x = 140$ mph

Check:

$4(140 - 20) = 3(140 + 20)$

$480 = 480$

Appendix F

Greek Alphabet

English / Pronunciation	Uppercase	Lowercase
ALPHA (AL-fuh)	A	α
BETA (BAY-tuh)	B	β
GAMMA (GAM-uh)	Γ	γ
DELTA (DEL-tuh)	Δ	δ
EPSILON (EP-sil-on)	E	ε
ZETA (ZAY-tuh)	Z	ζ
ETA (AY-tuh)	H	η
THETA (THAY-tuh)	Θ	θ
IOTA (eye-OH-tuh)	I	ι
KAPPA (KAP-uh)	K	κ
LAMBDA (LAM-duh)	Λ	λ
MU (MYOO)	M	μ
NU (NOO)	N	ν
XI (KS-EYE)	Ξ	ξ
OMICRON (OM-i-KRON)	O	o
PI (PIE)	Π	π
RHO (ROW)	P	ρ
SIGMA (SIG-muh)	Σ	σ
TAU (TAU)	T	τ
UPSILON (OOP-si-LON)	Υ	υ
PHI (FEE)	Φ	φ
CHI (K-EYE)	X	χ
PSI (SIGH)	Ψ	ψ
OMEGA (oh-MAY-guh)	Ω	ω

Appendix G

Glossary

Addition - the process of finding the number that is equal to two or more numbers grouped together

Algebra - the branch of mathematics in which the operations of arithmetic are generalized by the use of letters to represent quantities. Algebra is the abstraction of arithmetic.

Arithmetic - the study of quantity, especially as the result of combining numbers and is formally defined as the science of number and computation

Arithmetic sequence - a number series in which each term may be obtained from the preceding one by adding a constant called the common difference.

Base - The size of the groupings in a numeration system; for example, base ten (0-9)

Binary number system - a number system whose base is 2; the binary number system only uses the symbols 0 and 1

Cancelation - the act of dividing like factors out of the numerator and denominator of a fraction.

Complex fraction - a fraction that contains one or more fractions in its numerator, in its denominator, or both

Coordinate system - any method by which a number or set of numbers is used to represent a point, line, or geometric object. The set of numbers are called coordinates of the point.

Decimal - any proper fraction in which the denominator is some power of ten

Digit - one of the symbols of a number system

Division - the process of determining the number of times a given number contains another number. Division is the inverse of multiplication or repeated subtraction.

Exponent - a number placed at the right of and above a number or a symbol. The value assigned to the symbol with this exponent is called a power and indicates the power taken or how many times the number or symbol is multiplied by itself.

Extraneous solution - a number obtained in the process of solving an equation, which is not a root of the equation. It is generally introduced by squaring or clearing denominators in the original equation.

Fraction - a number that can represent part of a whole (part-to-whole ratio); the quotient of two rational numbers; a part of a unit or an indicated quotient of one number divided by another; or a numerical representation indicating the quotient of two numbers.

Fractional exponent - an exponent expressed as fraction that indicates a root of an expression. The numerator indicates the power to which the base is to be raised, and the denominator, the root which is to be extracted of that power.

Geometric sequence - a number series in which each term may be obtained from the preceding one by multiplying by a fixed number called the common ratio.

Greatest Common Divisor (GCD) - the greatest number or expression that is a factor of two or more numbers or expressions

Hexadecimal number system - a number system whose base is 16. The letters A, B, C, D, E, and F represent the hexadecimal digits

Improper fraction - a fraction in which the numerator equals or is greater than the denominator or has a value greater than one

Least Common Denominator (LCD) - the least common multiple of the denominators of a set of fractions. It is the smallest positive integer that is a multiple of the denominators.

Least Common Multiple (LCM) - the smallest number that is divisible by each member of a set of numbers

Linear equation - an equation that may be expressed in the form $ax + by = c$, where a, b, and c are any real numbers and $a \neq 0$

Logarithm - the exponent, n, to which the base b must be raised to equal a, written as $\log_b a = n$.

Mathematics - the collective name applied to all those sciences in which operations in logic are used to study the relationship between quantity, space, time, and magnitude. Mathematics employs a special kind of language using symbols, numerals, and letters.

Multiplication - simplified repeated addition

Number - a concept of quantity which could be a single unit or a collection of units; the value of the quantity is represented by a symbol called a numeral

Number line - a line, each point of which is associated with the number that gives its distance from a fixed point (the origin or zero) on the line

Numerals - the symbols of a numeration system

Numeration - the process of writing or stating numbers in their natural order

Mixed fraction - sometimes referred to as a mixed number or mixed expression, is a whole number and a fraction taken together

Order of operations - the process of following established procedures for the order of arithmetic operations

Percent - a certain rate per one hundred or a part of a whole expressed in hundredths

Percentage - the result obtained by multiplying a number by a percent (the percentage equals the rate times the base)

Perpendicular - lines that form 90 degree right angles.

Place value - a numeration system in which a real number is represented by an ordered set of characters where the value of a character depends on its position

Plane - a flat (no height) two dimensional surface, which contains at least three non-collinear (not on the same line) points and extends to infinity in all directions.

Proper fraction - a fraction in which the numerator is less than the denominator

Proportion - the relation between things (or parts of things) with respect to their comparative quantity, magnitude, or degree. Mathematically, a proportion is a statement of equality between two ratios.

Radical - an expression used to indicate the root of a number

Ratio - a comparison of two like numbers or quantities, and is the relative magnitude (size) of two quantities

Reciprocal - the number whose product with a given number is equal to one. Reciprocal quantities are any two quantities which produce one when multiplied together.

Sequence - a set of numbers in which each one is related in a definite way to the number that precedes it

Series - an indefinite number of terms succeeding one another, each of which is derived from one or more of the preceding by a fixed law, called the law of the series; as, an arithmetical series; a geometrical series.

Subtraction - the process of finding a quantity which when added to one of two given quantities will give the other. Subtraction is the inverse (opposite in order) of addition.

Summation - the act or process of forming a sum

Variation - the relationship between two quantities in which a change in one quantity results in a change in the other quantity.

Zero - As a digit, 0 is used as a placeholder in numeral systems using positional notation; zero also acts as the additive identity of the integers, real numbers, and other algebraic structures

Index

E

F

G

H

I

J

L

M

N

O

P

Q

R

S

T

V

W

Z

Made in the USA
Monee, IL
08 November 2026

67378645R00079